GREENHOUSE GARDENING FOR THE ABSOLUTE BEGINNER

HOW TO BUILD AN INDOOR FOOD OASIS AND GROW YOUR OWN FRESH PRODUCE YEAR-ROUND, NO MATTER YOUR SPACE, SKILLS, OR SEASON

JADEN "JR" RIVERS

ALL WE NEED PUBLISHING

Copyright © 2025 - All rights reserved.

This book is copyright protected. No portion may be copied, reproduced, distributed, quoted, or adapted in any form without written permission from the author or publisher, except as permitted under U.S. copyright law.

This book is for personal use only. Commercial use, resale, or distribution is prohibited.

Disclaimer Notice:

This publication is intended to provide accurate and authoritative information on the subject matter covered. It is sold with the understanding that neither the author nor the publisher is engaged in rendering legal, medical, financial, or other professional services or advice. Always consult a qualified professional before acting on any information contained herein.

The content is for educational and entertainment purposes only. Every effort has been made to present accurate, up-to-date, and reliable information. However, no warranties of any kind are expressed or implied.

The author and publisher make no representations or warranties regarding the accuracy or completeness of the contents. They specifically disclaim any implied warranties of merchantability or fitness for a particular purpose. No warranty may be created or extended by sales representatives or written sales materials.

The author and publisher assume no responsibility for your use of the material. You are solely responsible for your choices, actions, and outcomes. The advice and strategies contained herein may not be suitable for your situation. You should consult with a professional when appropriate.

Neither the publisher nor the author shall be liable for any loss of profit or other damages, including but not limited to special, incidental, consequential, personal, or indirect losses. This includes, but is not limited to, any errors, omissions, or inaccuracies contained within this book.

ISBN:

978-1-962344-03-6 (Ebook)

978-1-962344-04-3 (Paperback)

978-1-962344-05-0 (Hardcover)

LCCN:

Library of Congress Control Number: 2025909085

CONTENTS

Introduction: Through the Greenhouse Glass   vii

1. **GREENHOUSE 101**   1
   What It Is, Why It Works, and How to Make It Yours

   How Greenhouses Really Work: The Secret to Year-Round Gardening Bliss   2
   Greenhouse Myths: Let's Clear the Foggy Glass   4
   Get Your Greenhouse Goals On Paper   6

2. **CHOOSING YOUR GREENHOUSE**   13
   Types, Materials, and What to Know

   Greenhouse Types, Big and Small: Find Your Perfect Fit   13
   Sifting Through Options & Gaining Clarity   22
   Building Blocks: Choosing the Right Materials   25
   How to Build Smart: Choices That Fit Your Skills & Budget   29
   Choosing the Right Location: Sunlight, Space & Structure   35

3. **INSIDE THE GREENHOUSE**   40
   How to Design, Organize, and Equip Your Grow Space

   Design Your Dream Greenhouse: Layout & Aesthetics   40
   Master Tool Guide   43
   Where Function Meets Beauty   49
   Setting a Realistic Budget: Cost-Saving Tips & Tricks   51

4. **CLIMATE CONTROL**   55
   Creating the Ideal Environment

   Greenhouse Monitoring Tools: Your Early Warning System   56
   Regional Considerations: A Quick Note   59
   Temperature Control: The Goldilocks Zone   60
   Humidity Management: Keeping Things Just Steamy Enough   65
   Ventilation: Keeping the Air Moving   66
   Lighting Solutions: Because Your Plants Deserve a Spotlight   69

5. **WATERING WISDOM**   76
   Tools, Techniques, and Timing for a Thriving Greenhouse

   Understanding Individual Plant Needs   76
   The Secret Language of Thirsty Plants   77

Watering Like a Pro: Timing, Technique & Tips	80
Fancy Upgrades to Your Watering Toolkit	81
Automated Irrigation: Set It, Forget It, Let It Flow	82
Drip Irrigation: The Lazy Gardener's Best Friend	84
Rainwater Harvesting: Because Free Water is the Best Water	87
Troubleshooting: When Things Go Wrong (And They Will)	90

6. HOW TO BUILD YOUR GREENHOUSE — 93
Foundations, Framing, and Finishing Touches

Choosing Your Foundation Type	93
The Level-Headed Approach to Site Prep	96
The Great Assembly: Putting It All Together	96
Doors and Windows: Getting Your Greenhouse to Open Up	98
Safety First (Because Nobody Wants a Greenhouse Mishap)	100

7. SOIL AND FERTILIZATION — 105
Laying the Groundwork for Growth

Understanding Soil: It's Not Just Dirt, It's Dinner!	105
Fertilizers: The Protein Shakes of the Plant World	113
Mulch: Your Soil's Favorite Blanket	120
Composting: Where Kitchen Scraps Go to Graduate	121

8. PLANTING YOUR FIRST PLANTS — 127
Choosing, Planting, and Growing with Solid Strategy

Getting to Know Your Plants	128
Timing Is Everything (But Don't Stress Too Much About It)	133
The Planting Process	136
Advanced Techniques: Level Up Your Greenhouse Game	140

9. HARVESTING AND STORAGE — 146
Making the Most of Your Bounty

Knowing When to Harvest: How to Tell It's Go Time	147
Tools of the Trade: Your Harvesting Kit	151
The Gentle Art of Gathering	152
Storage Solutions: Keeping the Good Times Rolling	154
Preserving Your Harvest	156
Troubleshooting Guide: When Things Go Sideways	159
Why Preserve?	160

10. PEST AND DISEASE MANAGEMENT ... 164
    Protecting Your Green Paradise

    Creating a Pest Scouting Routine: Your First Line of Defense ... 164
    Meet the Usual Suspects: Common Greenhouse Pests ... 167
    Natural Pest Control: Your Green Defense Arsenal ... 170
    Plant Diseases: When Your Plants Need a Doctor ... 173
    Pest Prevention: Your Proactive Protection Plan ... 175

11. MAINTENANCE AND TROUBLESHOOTING ... 179
    Your Greenhouse's Health & Wellness Plan

    Maintenance to Keep Your Plant Spa Spic-And-Span ... 179
    Seasonal Shifts: Helping Your Greenhouse Adapt ... 181
    The Detective's Guide to Troubleshooting ... 182

12. BLOOMING BEYOND THE GREENHOUSE ... 191
    How Plants Grow People & Communities

    The Power of Reflection: Your Garden's Story ... 192
    Your Greenhouse, Your Sanctuary ... 193
    Building Your Garden Community ... 194
    Setting New Goals: Keep Growing! ... 195

    Conclusion: You Did It! ... 199

    A Note from the Publisher ... 202
    More From All We Need Publishing ... 204
    References ... 205

INTRODUCTION: THROUGH THE GREENHOUSE GLASS

Stepping into a greenhouse for the first time is a bit like stumbling through the wardrobe into Narnia. Except instead of talking lions, you find talking tomatoes!

Okay, maybe they don't talk. But *if they could,* they'd tell you: the greenhouse life is *the good life*.

My first greenhouse visit was quite the revelation. There I was, escaping the winter blues, when BAM—this humid hug of warmth wrapped around me like a tropical vacation for my sinuses. The air was thick with that earthy perfume that makes plant people go weak in the knees—you know, that intoxicating mix of soil, greenery, and pure possibility. Rows of plants were living their best lives, completely ignoring the fact that Mother Nature was throwing a frozen tantrum outside, and the rest of the garden looked like a popsicle convention.

Talk about having your kale and eating it too … in January! (Take that, winter!)

Of course, my journey from greenhouse groupie to actually growing things wasn't exactly smooth sailing. Let's just say it was … educational. I remember the days, armed with nothing but enthusiasm and a watering can, playing a game of "Too Much or Too Little?" with my poor plants. Spoiler

INTRODUCTION: THROUGH THE GREENHOUSE GLASS

alert: Turns out, there is such a thing as both of those. But hey, every wilted leaf is just a lesson in disguise, right?

Looking back, I'll tell you: a relationship with a garden can seem volatile in the beginning, but ultimately, a garden is a forgiving teacher if you stick with it. Even if its feedback system is basically a game of charades stretched out over years. Gradually, patiently, sometimes losing-my-patiently, along with lots of laughing at myself and learning the hard way ... I began to understand the silent language of my plants. And ultimately, my errors contributed to my garden getting better each year.

And guess what else? All those facepalm moments of mine (and others) are about to become your fast-track to success. Because nothing says "welcome to gardening" quite like learning from someone else's well-fertilized errors.

Now, it's true, making your own mistakes is an important, beautiful part of the process. Rest assured, you'll get plenty of your own to tell others about. But this book can be your VIP pass to skip the "what was I thinking?" phase of greenhouse gardening. Consider it like having a slightly over-caffeinated friend (that's me!) who's already made plenty of mistakes and is excited to share the solutions.

We're talking practical, real-world, roll-up-your-sleeves advice that works whether you've got a reputation as the local plant hospice or you're ready to level up your garden game. And while I can't promise you a magical wardrobe with talking tomatoes, I can definitely help you *grow* tomatoes while your neighbors are still chipping ice off their windshields. Which is still pretty magical.

Let's talk perks. Imagine your kitchen as a year-round farmers' market, except the produce section is right in your backyard. Fresh herbs in February? Check. Homegrown salads while the world is shoveling snow? Double check.

Plus, you get to craft a beautiful, peaceful space where stress can be left at the door—unless you're like me and accidentally leave the door open on a windy day. (Pro tip: Don't do that.)

Also plus, growing your own food is like giving the planet a high-five while saving money on groceries. Not to mention, it's great for bragging rights. That's what I call a win-win-win.

INTRODUCTION: THROUGH THE GREENHOUSE GLASS

You might be thinking: "This sounds amazing, but also terrifying and probably expensive." Deep breath, friend. Greenhouses are like jeans—they come in all sizes and prices, and finding the right fit is half the fun. Whether you're working with a champagne budget or more of a recycled-soda-bottles situation, we'll figure it out together. There's a greenhouse solution with your name on it—and I promise we won't have to raid your retirement fund to find it. (Though if anyone asks, we'll call it an "investment in botanical infrastructure.")

And complexity? Please. If I can figure it out after treating my seedlings like they were preparing for life as aquatic species (aka overwatering), you've got this in the bag. If you can follow a recipe or assemble flat-pack furniture, you're already ahead of the game. (And hey, if you can't do either, don't worry—I've got your back!)

As we dive into these pages, we'll start with the basics—like picking a greenhouse that doesn't require selling a kidney. Then we'll move on to the fun stuff: setting it up, filling it with green friends, and keeping them alive (arguably important). And then (drumroll) the harvest! I'll walk you through everything from "Help, why is everything wilting?" to "I have too many tomatoes!" (It's a thing).

Whether the space you're working with is practically a postage stamp, or you have a little backyard to dream in, it's an exciting adventure. Get pumped! You basically get to build your own ecosystem. Pretty cool, right?

Here's my suggestion: grab a notebook, make some coffee, and get ready to scribble in the margins. This isn't some dusty textbook—it's more like your greenhouse's birth certificate, instruction manual, and diary all rolled into one. So use it! Get some potting soil on it. Trust me, you'll want to remember that brilliant idea you had at 2 AM about vertical cucumber growing. (Those are always the best ideas, aren't they?)

Remember, this is *your* adventure, and I'm here to help you take it. Whether you dream of a mini Eden or simply want to grow some decent lettuce, we're in this together. Let's make it happen.

And hey, who knows? By this time next year, you'll probably be the one dishing out growing tips at neighborhood barbecues, giving tours of your winter wonderland, watching others' eyes light up as they step into your very own slice of perpetual summer. Now, wouldn't that be something?

INTRODUCTION: THROUGH THE GREENHOUSE GLASS

Maybe by then, you will have gotten to know your plants so well that you actually *can* hear the tomatoes' secrets.

As you embark, remind yourself that every master gardener started somewhere, probably killing a cactus or two along the way (ask me how I know). Celebrate where you are right now, and the fact that you're taking this step. Celebrate every little wobbly sprout and tiny bloom along the way. Embrace the whole glorious process, mistakes and all.

Oh, and in case you haven't noticed already, here's a really great part of that process: you're becoming part of a truly awesome community. Greenhouse gardeners are a delightfully eclectic group, ranging from the tech-loving hydroponics fanatic to the heritage seed saving traditionalist who can trace a tomato's family tree back further than their own. What connects us at the root level is our shared passion for growing: both our plants and ourselves.

So welcome to the jungle, friend. Like well-paired companion plants cozied up in a raised bed, we grow better together. Our thumbs get greener through shared wisdom, collective failures (oh, the stories we could tell), and getting our hands in the dirt with determination, again and again. So let's roll up our sleeves, get our hands dirty, and create something extraordinary together.

1

GREENHOUSE 101

WHAT IT IS, WHY IT WORKS, AND HOW TO MAKE IT YOURS

You might be standing at the edge of your backyard, clutching a steaming mug of coffee, staring with determination at that little patch you've always known has potential.

Or maybe you're staring at the ceiling and having an existential produce crisis. You know, where your mind fills with visions of sad, off-season tomatoes at the grocery store and you wonder ... I wonder if I can grow some of my own that don't taste like distant memories of actual tomatoes ...

I say: *yes, you can!*

Whatever your reason, you're here, and that's a great first step. And can I just say? That little spark of "maybe I could..." in your mind? That's how all great gardening adventures begin. No grand transformation montage required. (Although, if you want to play epic music and dance while you build this thing, I absolutely support that choice).

While it all might feel a bit mysterious now, by the end of this chapter, you'll see it's more like learning to ride a bike—with a few wobbles and a bit of practice, you'll be on your way to creating your very own verdant paradise. Just keep getting up and pedaling.

HOW GREENHOUSES REALLY WORK: THE SECRET TO YEAR-ROUND GARDENING BLISS

Ah, greenhouse gardening ... where winter is optional and the growing season never ends! In a greenhouse, you're not just extending the growing season; you're bending it to fit your lifestyle.

Think about gathering armfuls of fresh greens when the mercury dips below freezing, or treating yourself to homegrown peppers while winter storms rage outside. Seems too good to be true, but it's true.

So, you want some of that greenhouse magic in your life? Well friend, it's yours for the having! But it ain't magic. It's science with a party hat on, and it's about to become your new secret weapon.

Ever heard scientists talk about the greenhouse effect? You're essentially creating your own version of that! But unlike its global cousin, this one's actually working in our favor.

So, your greenhouse is basically running a clever little trick on nature. Sunlight passes through the transparent walls, transforms into heat when it hits objects inside, and then—plot twist!—the heat gets trapped in there, since it's less able to pass back through the glass. This creates your own personal bubble of extended growing season.

Abra cadabra. Thanks, science!

Think of your greenhouse as a cozy sweater for your garden. Not just that, it's protecting your plants from *all* of nature's mood swings—those nasty frosts, scorching heat waves, and pesky critters who think your vegetables are an all-you-can-eat buffet. (Sorry, rabbits, this is a members-only club!)

This is where it gets really fun: You become the director of your own little climate production.

Too hot? Open some vents and roll out some shade cloth like you're unfurling a beach umbrella.

Too cold? Your greenhouse traps that precious heat like a thermal sleeping bag.

Too cloudy? Supplement with grow lights and pretend you're a plant DJ spinning some solar beats. (Okay, got a little carried away with that one, but you get the idea!)

The point here is to zoom in on the power of greenhouses, which lies in their ability to offer environmental control that outdoor gardens simply can't match. You gain access to nature's override button. With simple tools like shade cloths and vents, you're suddenly the ruler of temperature, light, and humidity in your mighty glass kingdom.

With the proper setup, you can maintain a consistent environment where plants flourish regardless of what's happening outside. Beyond steady harvests, this consistency also grants you the superpower to experiment with plants that otherwise wouldn't survive your local climate. Look at you, you botanical rebel—defying local growing limitations.

Let's talk about what this means for your kitchen. Instead of relying on those sad, jet-lagged vegetables that have traveled farther than your last vacation, you're growing your own supply of five-star produce. There's something incredibly satisfying about serving a salad and being able to say, "Oh, this? Just picked it from my greenhouse!" (Try not to look too smug when you say it ... *but we both know it feels really good* ...)

So how do you achieve this yummy satisfaction?

Well, here's the first and most important key. It's very simple.

Just. Keep. Going.

Your greenhouse journey isn't about being perfect—it's about growing (pun intended) a little bit every day.

Take Margaret, for instance. Despite having a notorious brown thumb, she decided to try greenhouse gardening. Her first attempts were fraught with challenges—overwatering, under-lighting, and a mild infestation of aphids.

But with persistence, she turned her greenhouse into a lush oasis, her initial fears replaced with triumph. She connected with a local gardening group, shared her experiences, and received invaluable support. Today, she's growing far more than luscious plants. She's also growing confidence and community.

This story is not only possible; it's *typical* when there is perseverance. With simple goals and consistent, persistent effort towards them: it could be your story, too.

GREENHOUSE MYTHS: LET'S CLEAR THE FOGGY GLASS

Maybe you've heard some wild tales about greenhouse gardening. Maybe you think they're high-maintenance tropical resorts that you need a PhD in climate science to operate. (Spoiler alert: They're not, and you don't.)

Sometimes you gotta address the nagging questions before moving on. I get it. So, let's shine some light through these cloudy misconceptions, shall we?

Myth #1: "Only expert gardeners can handle greenhouses."

Truth: Listen. You can absolutely handle a greenhouse. It doesn't have to be perfect, and you'll learn along the way. Consider it one big experiment! Think of a greenhouse as your garden's training wheels—it actually makes growing *easier* because you control the environment.

Once you have a basic space, you can try different things inside of it. You don't have to build a mansion, just a tiny home. Start with some easy-going tenants like herbs or leafy greens, which can build confidence. Then expand from there.

Myth #2: "Winter growing requires expensive heating systems."

Truth: Your plants need warmth, not a tropical resort. Simple, budget-friendly solutions like thermal mass (fancy talk for water barrels or stones that soak up heat during the day and release it at night), good insulation, and strategically timing your growing seasons can keep things cozy without hemorrhaging money.

Myth #3: "Summer turns greenhouses into salad saunas."

Truth: Yes, an unventilated greenhouse can turn into nature's sauna. But basic ventilation (roof vents, side panels, or fans) and shade cloth work like your greenhouse's AC system. Think of it as giving your plants a nice sun hat and a cool breeze. Cost? A fraction of what you'd spend replacing heat-stressed plants.

Myth #4: "Greenhouses require constant attention—it's basically a part-time job."

Truth: Your greenhouse won't throw a tantrum if you skip a day. With basic automation (like timers for watering, temperature-controlled vents) and smart planning, you can create a low-maintenance setup that works with your schedule. Think more "set it and forget it" and less "helicopter gardening."

Myth #5: "You need tons of space for a greenhouse to be worth it."

Truth: A productive greenhouse is like a good book—size doesn't determine value. Even a modest greenhouse can be surprisingly productive. 6x8 structures can provide a steady stream of fresh veggies. It's about maximizing space, not maximizing space requirements.

Myth #6: "Greenhouses are too expensive for the average gardener."

Truth: While you can spend as much as a small car on a fancy setup, you absolutely don't need to. From DIY builds to budget-friendly kits, there's a greenhouse for every wallet. Plus, calculate the savings on groceries—those $5 organic herbs add up fast!

Myth #7: "Plants in greenhouses are more susceptible to disease."

Truth: Actually, your greenhouse plants are in a bubble—protected from many outdoor nasties. With good ventilation, and smart spacing, you *can reduce the risk* of pests and diseases. With a little due diligence, disease can be less of an issue than in outdoor gardens. Though fair warning: once your plants start thriving, you might develop a case of "just one more plant" syndrome. (Still searching for a cure for that one.)

Myth #8: "Greenhouses eliminate all pests."

Truth: Speaking of bubbles, sorry to burst this one. Yes, pests can still sneak in. *However,* here's the good news: It's easier to spot and deal with them since your plants are in one convenient spot. No more playing "Where's Waldo?" with aphids across your entire backyard.

Myth #9: "You can only grow certain types of plants in a greenhouse."

Truth: Your greenhouse is more flexible than a gymnast. While some plants might be happier than others, you'd be amazed at what you can grow. From ordinary tomatoes to exotic orchids, if you can dream it, you can probably grow it. Just maybe not a full-sized sequoia—let's keep things reasonable, folks.

The bottom line? Most greenhouse "problems" have simple, affordable solutions.

Final Thoughts

Remember, every expert gardener started as a beginner who decided to ignore the myths and give it a try (and probably many tries). Your greenhouse journey might include a few amusing mishaps, but that's part of the fun.

Besides, plants are surprisingly resilient—they've been growing for many years without our help. We're just giving them a cozy place to do their thing.

GET YOUR GREENHOUSE GOALS ON PAPER

Alright, garden dreamers, let's dream some dreams. Grab your favorite notebook (you know, the one you've been saving for "something special"—this is it!), or carve out some of that margin space.

Find Your Why

Let's begin your greenhouse adventure with a bit of introspection. It's time to ask yourself, "why are you really here?" Easy there, Socrates—not *THAT* deep ... I mean why are you *here*, reading a book on greenhouse gardening?

(Side note for you philosophical types, you'd be surprised what kind of profound revelations come to you in a garden!)

Why are you drawn to the idea of a *greenhouse*? It might be the allure of growing your own food and knowing exactly what goes into it. Imagine biting into a crisp cucumber that you nurtured from seed to harvest.

Maybe your motivation is rooted in sustainability, a desire to reduce your environmental footprint by growing your food more efficiently and helping to create a greener planet.

Or perhaps you're looking for a hobby that offers a break from the daily grind, a space where you can lose track of time and find a bit of peace. Pinpointing your primary goals will help shape every decision you make as you move forward.

Journal: Find Your Why

1. **Brainstorm reasons.** Write down 5 reasons you want to make a greenhouse. They can be ANY reason, including just wanting to prove to yourself that you can do it!
2. **Find your anchor.** Step back and take a look at the reasons you wrote down. If need be, revisit them another day with fresh eyes. Circle the reason that makes your heart do a little happy dance. You know, the one that makes you think, "Yes, *THAT'S* it!" This is your greenhouse North Star. Underline this one. It could be a couple of them.
3. **Make it real.** Here's where it gets fun—turn your "why" into an "I am" statement. Instead of "I want to grow food for my family," try "I am growing fresh, healthy food for my family." Feel that shift? That's you already becoming a greenhouse gardener! (Sneaky, right?) *You're doing it!* This is your anchor. Remembering why you're doing it will fuel you on days when this gets hard.

Your reasons may adapt as you go, but I suggest putting these somewhere you can revisit them often.

The Seeds of Something Greater

As you're jotting down greenhouse aspirations, here's a sneak peak into the future: a greenhouse tends to grow more than plants. You'll hear this theme echoed throughout the book—for good reason. I've seen it happen over and over again.

As individuals nurture their greenhouse gardens, they themselves inevitably grow something more profound in the process.

Consider this: each time you repair a ventilation problem or coax a diseased plant back to health, you're developing problem-solving abilities and confidence. You're developing community ties when you host the neighbors for your inaugural harvest dinner or exchange cultivation advice at the gardening shop. And those moments of quiet spent tending your crops? They're developing peace of mind and awareness.

I'm just saying, expect growth in many ways, and take note when you see it. Your journey with your greenhouse is going to sow some very wonderful seeds—in your garden and in your life.

Okay, fade out the inspirational music. Let's get practical and see what you are working with.

Working With What You've Got

Setting realistic expectations is the next step. Let's be honest, not all of us have the time or resources to cultivate an Eden by the end of the year. If you're juggling a job, family, and a social life, huge projects may feel overwhelming. But starting small can be just as rewarding. You could begin with a few pots of herbs or a raised bed of salad greens.

Consider your budget, too. Maybe you're eyeing that state-of-the-art greenhouse with bells and whistles, but if it's not within reach, plenty of budget-friendly options won't leave your wallet weeping. As you sketch out your plans, think about the time you can realistically devote to maintaining your green oasis.

Aligning your greenhouse goals with your lifestyle is equally important. If you have children, involving them in gardening activities can be educational and fun. For instance, you could spend a Saturday morning together planting seeds or picking ripe cherry tomatoes. My kids love spreading seeds, and when they finally see the sprouts come up, it's inspiring for everyone.

Let's get real about inviting a greenhouse into your life. Remember, working with what you've got isn't all about limitations—it's about getting creative.

Journal: Working With What You've Got

1. **Time detective work:** Track your typical week. Where are those golden pockets of potential garden time hiding? Maybe it's that 30-minute window while dinner's in the oven, or those weekend mornings before the kids wake up.
2. **Resource roundup:**
   - What's your "greenhouse fund" looking like? (Be honest—we're talking real numbers, not lottery-winning fantasies.)
   - What skills are already in your toolkit? (Even if it's just, "I successfully kept a cactus alive once.")
   - Who's your potential garden squad? (Family members, neighbors, that friend who posts plant photos all day.)
3. **Location scouting:** Write down every possible spot for your greenhouse, even the weird ones. Some great gardens start in unexpected places!
4. **Examine priorities:**
   - What are your priorities in your life? (e.g. "being an actively involved parent in my children's lives" ... your work week, home duties, caring for animals, plus any other activities that are important to you, like vacations or hobbies)
   - Consider the implied limitations and list your non-negotiables (like "must be able to maintain it in under 3 hours a week" or "must be able to leave for a week or two").
   - Now, find potential ways to incorporate greenhouse progress into your life while keeping priorities (involving your children in the process, listening to gardening podcasts during your work commute or while cooking meals).

Getting Deliciously Specific

Time to zoom in. You wrote down *why;* now, let's write down *what.* If you got this book, I assume you have some interest in gardening. Let's look around your life even more, finding things you already love to do and asking, how can a greenhouse enhance this?

For instance, if you love cooking, your greenhouse can become a source of fresh ingredients, adding a farm-to-table flair to your meals. Maybe you're into crafting. Consider growing flowers that can be dried and used in your projects. Your greenhouse can be an extension of who you are and what you love.

Journal: Getting Specific

1. **Create your greenhouse story.** How? Imagine your greenhouse and describe what you see. Here are some prompts to inspire you:
   - "Every morning, I walk into my greenhouse and ..."
   - "The first thing I want to grow is ... because ..."
   - "My favorite time to be in the greenhouse will be ... when I'm ..."
   - "I can't wait to share ... with ..."
   - "The moment I'm most looking forward to is ..."
2. Now, **get specific** with your greenhouse dreams. Be really descriptive!
   - What will you do with your harvests? (Making pasta sauce from home-grown tomatoes? Mixing fresh herb cocktails?)
   - Who will you share your garden bounty with?
   - What problems will this solve in your life? (No more sad grocery store herbs!)
   - How would your greenhouse make you feel? (Proud? Peaceful?)

Logging the Adventure

Alright, since you've been journaling a lot, I'll make mention of one important use of your journal: *tracking progress*. Tracking your progress can be both enlightening and motivating. Keep a simple journal where you jot down what you've planted, what works, and what doesn't.

You might note that your peppers thrived on a sunny shelf while your lettuce preferred the cooler corner. This record becomes a valuable resource, guiding your choices and helping you adjust your goals as you learn.

Flexibility is key, because gardening is as much about adapting as it is about planning. Sometimes, a plant will surprise you with unexpected vigor or stubbornness, and your goals might shift in response.

Keep an open mind and enjoy the learning process.

Check-In

Alright, fellow greenhouse gardener, time to celebrate, because you've already made great progress. Don't think so? Let's look at it this way.

By opening this book, and thinking through all these questions, you've actually done a lot to lay the groundwork for your greenhouse adventure.

Look at what you've accomplished already:

- You now understand how greenhouses actually work (that whole trapping-sunlight-like-a-heat-seeking-missile thing).
- You've busted through common greenhouse myths that might have held you back.
- You've started mapping your greenhouse motivations and dreams in concrete terms.
- You've begun thinking strategically about your resources, space, and lifestyle.
- You're developing a realistic view of what greenhouse gardening involves (the good, the challenging, and the absolutely worth it).

This isn't just idle daydreaming anymore—you're building the foundation for real greenhouse success. That's significant progress on your journey from curious plant-fan to confident greenhouse gardener.

Once you've celebrated that, get out your journaling pencil one more time, and take a shot at the following simple questions. They'll help you with the next chapter and the rest of your greenhouse.

Ready? Okay, write down your responses to the following:

- Have you identified your "why" for wanting a greenhouse? (That anchor that'll keep you motivated when nature throws its curveballs)
- Do you have a sense of the space you can realistically dedicate to this project? (Even just a ballpark range is great)
- Have you thought about what plants you'd love to have fresh year-round? (In your own personal grocery aisle)
- Have you considered how your greenhouse dreams align with your actual lifestyle? (Because even plant parents need work-life balance)

Having these answers in mind will help guide your decisions as we move forward.

Don't know them all yet? Totally cool. That's why I'm here! We'll figure it out together.

For now, jot down your initial thoughts. They'll develop naturally as we go.

Quick success tip: Keep these notes somewhere you can revisit them when decisions get tough. Your "why" is like your greenhouse's North Star—it'll guide you when you're choosing between options or feeling overwhelmed.

Remember: Even master gardeners started with nothing but questions and dreams. You're right where you need to be!

Wrap-Up

Alright, we've laid the groundwork for your greenhouse adventure! We explored how greenhouses work, cleared up some common myths, and wrote down your dreams to light the fire of motivation and guide your vision.

With a more solid understanding and goals in hand, you're ready to take the next step. In the next chapter, we'll dive into the exciting world of greenhouse types and materials to help you find the perfect fit for your space, style, and needs.

Ready to start turning these dreams into reality? Let's get started!

2

CHOOSING YOUR GREENHOUSE

TYPES, MATERIALS, AND WHAT TO KNOW

So you're ready to choose a greenhouse! How exciting! It can be a bit like being a kid in a candy store. It's amazing, but I mean, choosing between sour gummies and chocolate bars can be pretty tough. Or, in our case, polycarbonate versus glass panels.

The good news? There are about as many greenhouse styles as there are reasons to want one. The even better news? We're going to break down each option so you can choose with more confidence and less overwhelm.

Whether you're working with a backyard or a balcony, there's a greenhouse design tailored to your needs.

If you get overwhelmed, it's okay. Reexamine your space and options, dump some thoughts in a journal, rinse and repeat, and the answers will start to become clear.

GREENHOUSE TYPES, BIG AND SMALL: FIND YOUR PERFECT FIT

Let's break down your options, from the classic to the cosmic. There are both full-sized models and mini versions.

We'll make this like speed dating for greenhouses—focusing on the important stuff so you can make a more confident choice.

Full Greenhouse Types

Classic Gable-Style & A-Frame: The Mountain Lodge of Growing Spaces

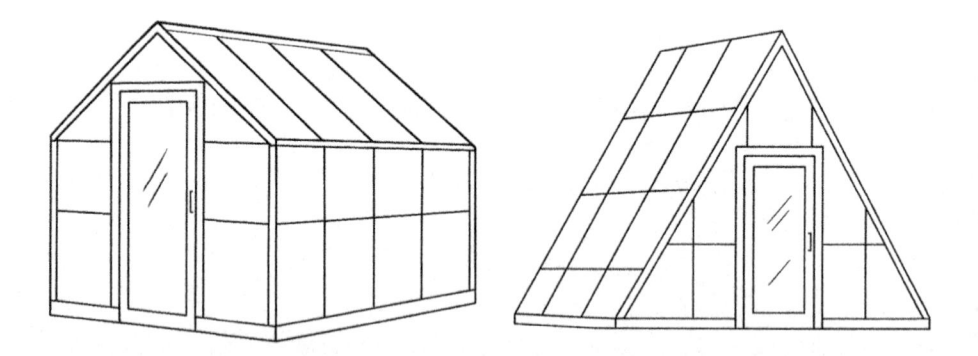

- **Conventional / Gable-style:** Features vertical sidewalls with a roof that slopes on both sides, forming a classic house-like shape. This design provides ample vertical space (great for installing features) and is ideal for hanging plants or tall crops. *Think: traditional house.*
- **A-frame:** Characterized by steeply pitched sides that meet at a peak, forming a triangular structure reminiscent of a tent. While it offers excellent snow and rain shedding capabilities, the sloped walls can limit usable interior space near the base. *Think: tent.*
- **Cost:** Medium-high
- **DIY-friendly?** Yes
- **Best for:**
    - Snowy regions (excellent snow shedding)
    - DIY builders (straightforward construction)
    - Maximizing vertical growing space (Gable)
    - Budget-conscious growers (A-Frame)
- **Watch for:**
    - Limited headroom at the edges in A-Frame designs
    - Potentially higher construction complexity for Gable structures
- **Pro tip**: Utilize the vertical space in Gable greenhouses for hanging planters or trellised vines. In A-Frames, consider built-in shelving along the sloped sides to maximize growing area.

CHOOSING YOUR GREENHOUSE

Geodesic Dome: The Futurist's Dream

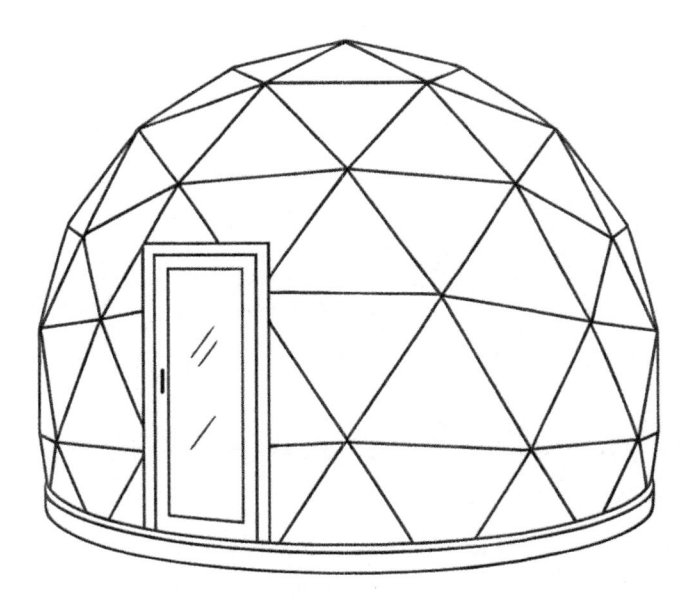

- **What:** Spherical structure made of triangular panels—your plants' personal planetarium. Don't let the spaceship vibes intimidate you though. Once it's up, maintenance is surprisingly easy.
- **Cost:** Premium
- **DIY-friendly?** Challenging
- **Best for:**
  - Wind-prone areas (superior aerodynamics)
  - Unique aesthetics
  - Maximum strength with minimal materials
  - Excellent light distribution most of the year (in winter, you may need a little extra supplemental lighting in higher latitudes.)
  - Energy-efficiency for heating and cooling, great air circulation
- **Watch for:** Complex assembly, specialized repairs.
- **Pro tip:** Assembly feels like solving a 3D puzzle, but the resulting strength-to-weight ratio makes these structures nearly indestructible. Your great-grandchildren might inherit this one!

Gothic Arch: The Cathedral of Cultivation

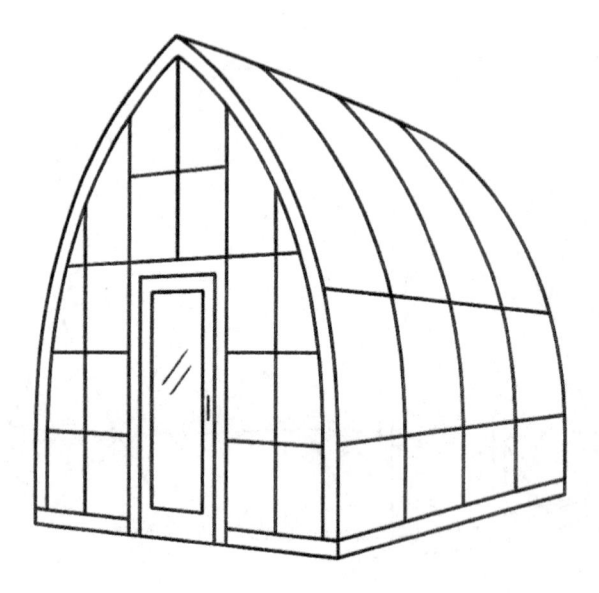

- **What:** Cathedral-style peaked arch design, like a cross between a hoop house and an A-Frame. You know, in case you can't decide between the two.
- **Cost:** High
- **DIY-friendly?** Moderate
- **Best for:**
    - Heavy snow areas (steep sides handle snow and rain well)
    - Maximum interior height—allowing for taller crops (even small trees) compared to hoop houses of similar width
    - Traditional gardens
    - Excellent moisture management: the curved roof sends condensation down the sides rather than onto plants
- **Watch for:** More complex construction.
- **Pro tip:** For DIY builders, using flexible PVC or metal conduit piping can simplify constructing the signature Gothic curve without needing custom framing.

CHOOSING YOUR GREENHOUSE

Hoop House: The Flexible Friend

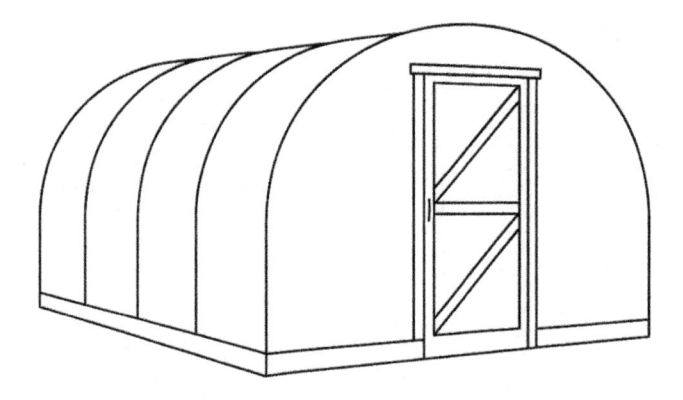

- **What:** Curved pipes covered with greenhouse plastic—the VW Beetle of greenhouses.
- **Cost:** Medium-low
- **DIY-friendly?** Yes
- **Best for:**
  - Budget-conscious growers
  - Temporary/seasonal growers
  - Large growing areas
  - A removable/portable option
- **Watch for:** Regular plastic replacement (every few years). Expect to replace the plastic covering every 4–5 years if UV-treated; non-treated plastics might only last 1–2 seasons.
- **Pro tip:** On the plus side, regular plastic replacement gives you a built-in deep cleaning schedule. For better insulation in colder climates, consider using double-layer plastic sheeting inflated with air.

Lean-To: The Space Optimizer

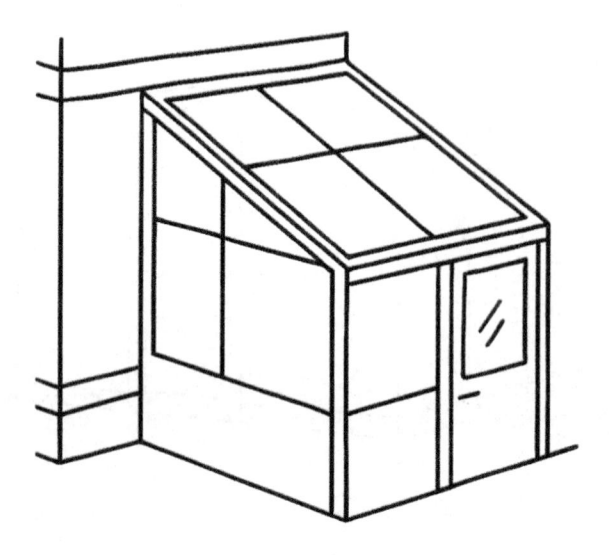

- **What:** Greenhouse attached to existing structure.
  - (*Abutting variation: Fully connects to a building, allowing indoor access. It's your home's "horticultural addition".*)
- **Cost:** Medium
- **DIY-friendly?** Moderate
- **Best for:**
  - Limited space and easy access
  - Using house heat (drawing thermal mass from the building)
  - Urban/suburban properties
- **Watch for:** Building permits, wall attachment issues. The attached building can cast shade, so be mindful of orientation and building height. For snowy regions: keep an eye on snow build-up between the house and the greenhouse to prevent structural stress.
- **Pro tip:** South-facing wall placement (in Northern Hemisphere) turns your house into a natural heat sink, keeping your plants cozy while reducing your heating costs.

CHOOSING YOUR GREENHOUSE

Mini Greenhouse Types

Mini Greenhouse: The Studio Apartment

- **What:** Small, freestanding (often portable) units with shelves — like a bookshelf with a greenhouse cover. Perfect for urban gardeners.
- **Cost:** Low
- **DIY-friendly?** Yes (assembly)
- **Best for:**
    - Balconies/patios, small yards, apartment dwellers
    - Seasonal protection
    - Starting seeds
    - Growing a curated collection of plants
- **Watch for:** Temperature fluctuations. Their small air volume means they can heat up very quickly in direct sun and cool down rapidly at night, so they'll need monitoring and ventilation.
- **Pro tip:** Place against a sunny wall for extra warmth and protection. Also, don't underestimate these little guys—mini greenhouses can kickstart a whole backyard garden if you plan smart!

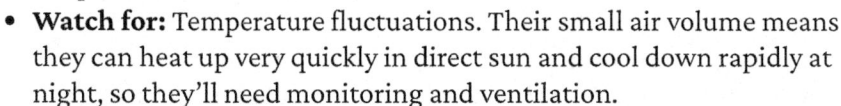

Cold Frame: The Gateway Growing Space

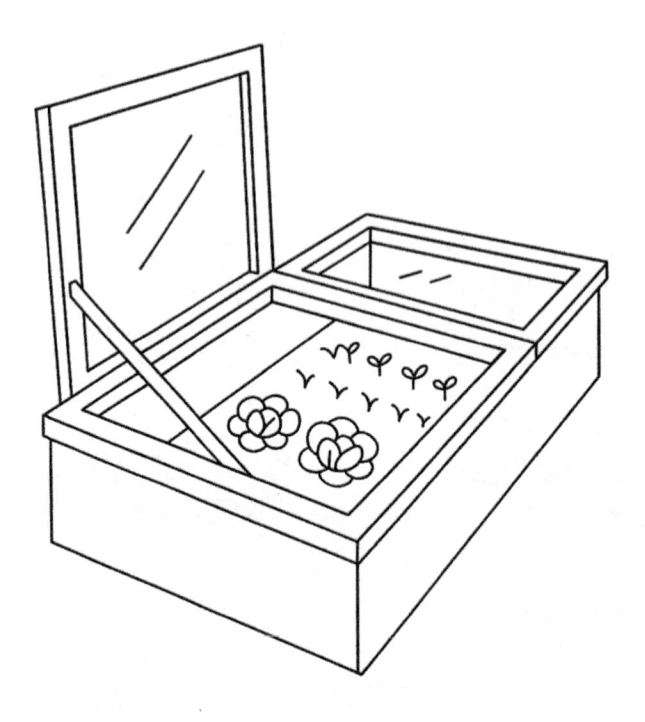

- **What:** A simple box with a transparent lid—imagine a miniature greenhouse that sits directly on the ground. Your greenhouse garden's starter home.
- **Cost:** Very low
- **DIY-friendly?** Very easy
- **Best for:**
  - Seed starting and hardening off plants
  - Season extension
  - Small space gardening
  - Budget-conscious growers
- **Watch for:** Manual ventilation needed. Also, cold frames mainly rely on solar gain and heat held in the soil for warmth, offering good frost protection—but it's usually not enough warmth to overwinter tender plants without added insulation or heating cables.
- **Pro tip:** Old windows make great lids—just secure them well. Flying glass is nobody's friend!

Cloche: The Personal Plant Protector

- **What:** Individual plant covers—the OG social distancing tool for plants. Creates a protective microclimate.
- **Cost:** Minimal
- **DIY-friendly?** Yes
- **Best for:**
    - Protecting individual plants
    - Early/late season growing
    - Frost protection
    - Small-scale gardening
- **Watch for:** Limited protection, size restrictions.
- **Pro tip:** Look for versions with ventilation options—plants need to breathe too. For DIY-ers: old milk jugs or large soda bottles can be repurposed into quick DIY cloches—cheap, cheerful, and effective.

Vertical Tower: The Space Revolutionary

- **What:** Vertical growing systems with greenhouse covering, like a high-rise for herbs.
- **Cost:** Medium-low
- **DIY-friendly?** Moderate
- **Best for:**
    - Limited ground space
    - Seed starting, herb gardens
    - Urban environments
    - Efficient use of light and space
- **Watch for:** Watering challenges, limited plant varieties.
- **Pro tip:** Often used on patios or indoors, or can be covered. Often work best with integrated hydroponic, aeroponic, or drip irrigation setups. Hand-watering these can be like playing Twister.

BEYOND BASICS: OTHER GREENHOUSE SHAPES TO EXPLORE

Here are a few other greenhouses you can look up for some research fun:

- **Pit Greenhouse (Walipini):** Dug into the ground to use the earth's natural insulation—great for extreme climates and harsh winters.
- **Quonset Hut**: Curved metal frame, built tough. A sturdier hoop house with excellent wind resistance. Ideal for large-scale growing.
- **Dutch Light:** Tall, elegant, and a champion at trapping weak northern sunlight. Its steep roof and large glass panes maximizes low-angle winter sunlight common in cloudy, northern climates.
- Explore **creative forms** like: flat arch, dome, tri-penta, sawtooth, skillion, uneven span, ridge & furrow, shade house, and igloo styles.

And there you have it—your grand tour of greenhouse types, from the classic A-frame to the underground Walipini and everything in between.

Whether you're dreaming big with a dome or starting small with a cold frame, there's a structure out there that'll suit your space, your style, and your growing goals.

SIFTING THROUGH OPTIONS & GAINING CLARITY

Now that we've had our whirlwind greenhouse speed date—complete with charming A-frames, space-age domes, and cozy cold frames—you might be wondering how to actually pick one. Not to mention all the other decisions you'll encounter on your great greenhouse adventure.

The next few sections are here to help you start clarifying your thought process—and maybe even turn a few question marks into next steps. Chances are, you've still got plenty of lingering questions before picking your match. Starting with ...

What's This Going to Cost?

The first thing you might be wondering after your speed date is: *But how much is this love affair actually going to cost me?*

Great question. And here's the honest dirt: I thought about mapping out some ballpark price ranges. Had them all lined up.

But in the end, I decided to stay in general zones. Why? Because greenhouse costs are slippery little things. They vary depending on your location, the time of year, the materials you use, whether you go full DIY or buy a kit, and whether your local lumberyard is running a sale or woke up grumpy.

There are low-end and high-end versions of all the above types. And, if you're looking for professional installation, that adds considerably to cost.

Not to mention, prices can fluctuate. Some things (like gardening basics) are timeless—other things (like exact prices) are not.

In short: variables abound.

It's still not *overly* complicated. The above list does give you a *general* idea of what kind of green you'll need for each type of greenhouse.

For specifics, it's best to do a little more digging online and asking around.

Below are some simple online searches you can do to get a good basic idea. It's good to insert your current year to get the most up-to-date results.

- "Greenhouse prices by type [insert year]"
- "Cost to build a DIY hoop house [year]"
- "Greenhouse kit prices by size [year]"
- "How much does a small greenhouse cost [year]"
- "Greenhouse materials cost breakdown DIY [year]"

Here's the good news: there's an option out there to fit your space, skill level, and wallet without wilting your enthusiasm.

Later in this chapter, we'll talk about budgeting, specific material details, and how to be creative and resourceful when building your greenhouse.

Narrowing Down: *Climate, Space, and Choosing with Confidence*

Speaking of variables, your local climate and available space play important roles in selecting the right greenhouse. For instance, if you live in a windy area, you'll need a sturdy structure with a strong frame to prevent it from becoming an unplanned kite.

A few ideas:

- Snowy region? A-frame or Gothic arch is your friend
- Windy area? Geodesic dome laughs at gusts
- Limited space? Lean-to or vertical mini greenhouse
- Tight budget? Start with a hoop house or a "mini" version (like a cold frame)

Considerations like this can help you steer more in the right direction. The more information you have about your specific situation, and the more you find what you like, the better you can make a decision that fits you.

But even with a clearer path, it's totally normal to glance at the map and still feel a little lost sometimes. We've got plenty of learning left to do on the road ahead, but it's good to look at this now before more options are in front of us.

Sometimes, the more options you explore, the more questions pop up. But that's actually a *good* thing. Each question brings you closer to the greenhouse that truly fits your life, because it makes you look at things and learn.

But in the moment? Yeah, your brain can feel like it's frozen and lost signal. If you get that feeling, don't worry—you're not alone. Right here at the start, let's address that first wave of something you may be feeling: *overwhelm*.

Here's the deal. It's easy to get swept up with the idea of finding the perfect greenhouse. It can feel like the more you try to chase after perfect, the fuzzier the picture gets of what that even is.

The reality is: the "perfect" greenhouse is the one that fits your space, budget, and dreams. Don't get too caught up in all the polished pictures online. The best greenhouse is the one that gets built and actually used!

If you find yourself too far in the weeds of choosing the "right" one, don't worry, I've definitely been there. Here's something simple I do.

Narrow down to your top choices. Then, walk away, sleep, look at your top picks again the next day, research the options, and simply pick one. (You can always be open to changing your mind a few chapters down the road).

Once you've narrowed down to similar options that all work for you, there usually is no "perfect" or "right" here. There are most likely multiple choices that will give you a good springboard into the greenhouse life.

I promise I get it. You'll be spending a lot of time in this space, so you want to get it right. You want to create a space where you can't wait to spend those early morning hours with your coffee and your plants. So yes! Do your due diligence.

But, if overthinking is slowing you down from making any choices, use one of my favorite phrases: *don't let the good be held hostage by the perfect.* Make a solid choice, then keep going. Because more often than not, you'll learn to love the one you've built—quirks and all. Plants don't care about perfection. They just need a space to grow, and so do you.

I know all the choices can be a lot to take in. I've been there, too. Along with every other greenhouse gardener! They all made it through to greenhouse glory, though, and you and I are going to rock this together. We'll sift through choices, breathe, and sift again. Your choice will become clearer, the more you learn.

So, ready to keep learning? Let's press on!

Planning For Growth: *Space Now, Room to Expand Later*

When planning your greenhouse dreams, think long-term. Consider how your greenhouse might evolve. Perhaps today it's a cozy nook for herbs, but tomorrow it could be a thriving haven for citrus trees or exotic orchids.

Another note on sizing. Whatever style you choose, remember this golden rule: Very few people have said, "Gosh, I wish I'd built a smaller greenhouse." If you *can* go bigger while staying within budget and space constraints, *do it.* Your future self (and those wild trailing plants) will thank you.

Modular designs offer the flexibility to expand as your gardening ambitions grow. Think of a modular greenhouse like gardening with LEGO bricks—it's built in standardized sections that can connect together. Your starter greenhouse might be 8x8 feet, but when you catch the herb garden bug (and trust me, fresh basil is a gateway plant), you can add another 8-foot section without rebuilding the whole thing.

As you plan for growth in all its forms, keep in mind that your foundation needs to support your greenhouse's future growth potential, not just today's setup.

Okay! Now that we've explored different types of greenhouses, what materials will you use to build them? That's what we'll cover next.

BUILDING BLOCKS: CHOOSING THE RIGHT MATERIALS

When talking materials, there are two main parts of the greenhouse to look at: the covering and the frame. "Covering" is a term used to describe the transparent panels, while the "frame" is the wood planks or pipe that holds it all together. The covering and frame are the skin and bones of your greenhouse, respectively.

First, let me break down your greenhouse frame options—the choices that'll quite literally support your gardening dreams.

Frame Materials

Plastic piping (PVC)

The beginner's best friend. Affordable, lightweight, and simple to assemble. Perfect for smaller structures and mild climates. Expect 5–10 years of service before UV exposure makes it brittle. The catch? They can wobble in strong winds. Pro tip: Spring for UV-resistant PVC—it's like sunscreen for your frame, and worth the slight up-charge for longer life.

Wood

The DIYer's delight. Strong, naturally insulating, aesthetically pleasing, and endlessly customizable. Cedar and redwood naturally resist rot and insects, making them ideal choices. The trade-off? Annual maintenance is non-negotiable. Think preservative coatings and solid drainage design to prevent moisture issues. But for the hands-on gardener, the effort pays off in durability and charm.

Metal piping

The long-term investment. Long-term durability, but doesn't have wood's insulating powers (especially aluminum). The premium choice splits into two camps: aluminum is your lightweight champion—won't rust, easy to move, perfect for smaller setups—while galvanized steel brings the muscle for serious snow loads and wider spans. Both laugh at rust, but steel's strength comes with extra weight and cost. Think aluminum for hobby houses, steel for forever homes. Pricier than plastic or wood, but they're marathoners, not sprinters—lasting decades with minimal (basically zero) maintenance.

Quick decision guide:

- Tight budget, mild weather → **PVC**
- Love DIY, don't mind maintenance → **Wood**
- Want set-it-and-forget-it durability → **Metal**

Remember: whatever you choose, proper anchoring is crucial—unless you're secretly building a greenhouse rocket ship. (Personally, I'd stick to growing plants.)

As far as frame material goes, you're basically looking for something that can withstand a beating from the elements. You don't want an unpleasant storm to leave your place resembling modern art instead of a greenhouse. After my own *ahem* unpleasant experiences, I tend to over-engineer everything. It's better to have it and not need it than to watch your greenhouse attempt to fly, right?

Next, let's look at your greenhouse covering options—the materials that'll turn sunlight into garden success. Each material has its sweet spot for clarity, durability, and cost.

Covering Materials

Covering options can be broadly categorized into three main groups: glass, rigid plastic sheeting (including polycarbonate, acrylic, and fiberglass), and flexible plastic films (such as polyethylene, EVA, and PVC). They all do the same basic job to different degrees and in different ways. Here's a breakdown to help you choose.

Glass

The classic choice: unbeatable clarity and decades of service if properly maintained. Tempered glass scoffs at scratches and holds heat like a champ. The catch? It's heavy, expensive, and one wayward baseball means replacement time. Perfect for permanent structures where looks matter, but demands serious framing support, and is harder to install. Pro tip: Double-paned glass gives you extra insulation but doubles down on the cost factor.

Polycarbonate

The practical powerhouse: lightweight, virtually unbreakable, and excellent insulation (especially twin-wall). Also more budget-friendly than glass or acrylic. Diffuses light beautifully to prevent plant scorching. UV-protected versions resist yellowing for 10+ years. Less crystal-clear than glass but makes up for it with better heat retention and impact resistance. Your best bet for most home greenhouses—like choosing an SUV over a sports car.

Acrylic

The middle-ground option: clearer than polycarbonate but tougher than glass. Great light transmission and decent insulation. Plus, it's durable and can be molded for curved arches. Here's the rub: it scratches easily and can get brittle with age. Choose it when you want glass-like clarity without glass-like fragility. Works best in areas without hail or extreme weather.

Fiberglass

The budget-friendly survivor: strong, lightweight, and surprisingly good at light diffusion. UV-protected panels can last 10–15 years before getting cloudy. Cheaper than the others but requires regular cleaning to prevent algae buildup, and it tends to turn yellow over time. Think of it as the working-class hero of greenhouse materials—not fancy, but gets the job done.

Polyethylene film

The budget-friendlier, temporary solution: lightweight, flexible, and surprisingly effective for short-term projects or seasonal extensions. Polyethylene film allows excellent light transmission and can be easily installed on simple frames, such as PVC hoops. UV-treated options can potentially last up to 4 years, while untreated films may need annual replacement. Although it lacks the insulation properties and durability of rigid panels, polyethylene film offers an economical way to start your greenhouse journey without breaking the bank. Consider it the "training wheels" of greenhouse coverings—perfect for testing the waters before investing in a more permanent structure.

CHOOSING YOUR GREENHOUSE

Quick decision guide:

- Want long-term durability + aesthetics → **Glass**
- Need practical performance + value → **Polycarbonate**
- Want glass-like clarity + lighter weight → **Acrylic**
- Need budget-friendly + decent performance → **Fiberglass**
- Starting small or temporary project → **Polyethylene film**

Remember: light transmission is great, but heat retention often matters more. When it comes to aesthetics, glass wins in clarity, but polycarbonate's diffused light promotes healthier plant growth by reducing hot spots. Choose based on your climate and comfort with maintenance, not just looks.

Maintenance-wise, glass requires regular cleaning to maintain its sparkle, while polycarbonate may need occasional repairs—often easily managed with a repair kit. Installation varies too: glass demands precision and possibly professional help, while polycarbonate and polyethylene are more DIY-friendly, saving you both time and money.

And if the decision feels overwhelming, just keep exploring—the more you learn, the clearer the answer will become. It's like one of those 3D hidden image posters—at first it's just a bunch of dots, but suddenly, BAM! The picture pops out at you. Trust the process, and soon you'll be seeing your perfect greenhouse in all its glory!

HOW TO BUILD SMART: CHOICES THAT FIT YOUR SKILLS & BUDGET

You don't need a construction degree or a closet full of cash to build a greenhouse that works for you—you just need a plan that fits *your* skills, *your* time, and *your* wallet.

Whether you're the type who loves sketching blueprints on napkins or you'd rather just follow a solid set of instructions with your coffee in hand, this section will walk you through some options.

And for those with a whole lot of dreams and not a whole lot of budget … we'll explore some creative solutions to spark your imagination.

DIY or DIMYBWAFKTDFY? *(Do it mostly yourself but with a few key things done for you)*

Choosing between full-on DIY versus a pre-fabricated DIY kit is like deciding whether to bake a cake from scratch or use a mix—it depends on your skills, time, and appetite for adventure. Either way, you get the satisfaction of having accomplished something yourself.

Full DIY gives you complete control and can save money (we're talking up to 30–50% less), especially if you're handy with tools and can source materials smartly. Plus, you can customize every detail, from the height of your benches to the exact placement of your vents. And you can find all kinds of plans online. But here's the reality check: DIY demands research, planning, and the confidence to troubleshoot when things inevitably go sideways.

Prefab kits, on the other hand, come with clear instructions, pre-drilled holes, and the peace of mind that someone's already worked out the engineering (no midnight panic about wind loads or foundation specs). They're like having a knowledgeable friend hold your hand through the whole process—though that friend charges for the service. Just look for a product with good reviews that fits your setup.

My advice? If you've built a deck or tackled similar projects, DIY could be your path to greenhouse glory. But if the phrase "load-bearing wall" makes you nervous, a kit will get you growing faster with fewer headaches.

Quick decision guide:

- Strong DIY skills + time to plan → **DIY / Build from scratch**
- Want customization + budget flexibility → **DIY**
- Need quick setup + clear instructions → **Kit**
- Value peace of mind + simplicity → **Kit**

Remember: Whether you choose full DIY, or a DIY kit, the goal is the same—creating a space where plants thrive and you enjoy spending time. Choose the path that lets you sleep at night—your plants don't care whether you built their home from scratch or assembled it from a kit. (Though they might appreciate you being less frazzled during planting time.)

Budget-Friendly Greenhouse Options: *Saving Without Compromise*

Creating your dream greenhouse doesn't have to empty your savings account. There's a world of affordable materials and designs waiting to be explored, including DIY greenhouse kits that offer a balance of cost and convenience.

As mentioned above, you sometimes pay for the convenience of prefab kits, but don't worry too much if you're short on both cash and time (and maybe DIY creativity). There are options for a variety of budgets.

The Savvy Scrounger's Guide

If you're feeling particularly thrifty and creative, consider using reclaimed materials. Old windows—once destined for the landfill—can be transformed into charming greenhouse walls that let sunlight flood in. Not only do these projects cut costs, but they also add a unique character to your greenhouse, making it truly one-of-a-kind.

When it comes to saving on construction, sourcing materials locally can significantly reduce expenses. Local suppliers often have surplus or discounted items that can be repurposed for your greenhouse needs.

Don't overlook community resources either. You'd be surprised how a casual conversation can lead to unexpected savings. It's a bit like a treasure hunt.

Remember that scene in Apollo 13 where they had to build a CO2 filter from spare parts? Embody that mojo. Here are some spots on your treasure map:

1. *Architectural Salvage Yards*: Gold mines for old windows and doors. Pro tip: Visit at month's end when they're clearing inventory.
2. *Construction Sites*: Many contractors are happy to part with "waste" materials like lumber offcuts or surplus roofing panels. Just ask politely and bring coffee.
3. *Online Marketplaces*: Set alerts for "greenhouse materials" or "used windows"—you'd be amazed at what people practically give away during spring cleaning.
4. *Neighbors*: Sometimes a neighbor or a local gardening club might have leftover materials from their own projects.

Money-Saving Master Moves

Create a materials spreadsheet before you start. List everything you think you'll need, from screws to soil. This helps you keep costs in check, while also making you aware of those surprise expenses that can pop up like weeds.

For larger costs, you can explore financing options. Some community programs offer grants or low-interest loans for sustainable projects, which could provide a bit of financial breathing room. Smart decisions upfront ensure smooth sailing during construction.

Buy in bulk with other gardeners—like a greenhouse materials potluck, minus the questionable casseroles.

Time your purchases around seasonal sales (think Black Friday for tools, end-of-season for materials).

The Resourceful Recycler

With a little bit of creative innovation, you'd be amazed what garden goodies you can create from everyday treasures, while trimming costs and adding a personal touch. Consider, for example, simple projects like building raised beds from reclaimed wood or forgotten pallets, or crafting a rainwater collection system using barrels and hoses.

Upcycling everyday items can turn a mundane task into an exciting project. An old shelf could become a multi-tiered plant stand, and those slightly bent nails in the garage might finally find their purpose. It's about seeing potential in the mundane and finding joy in the process of creation.

Here are a few examples of transforming trash into greenhouse treasure:

- Glass doors = instant greenhouse walls
- Old gutters = vertical growing systems
- Plastic bottles = mini cloches for seedlings

Just remember: avoid anything that's hosted mold or chemicals—your plants don't need that kind of drama in their lives.

Invest Where Important

Saving money is always a win, but there are parts of your greenhouse where cutting corners can cost you more in the long run. Think of it like building a car—you wouldn't skimp on the engine unless you're a pro mechanic (and even then, would you trust it on the highway?). Some areas need to be built with care and integrity to keep your greenhouse running smoothly and standing strong for years to come.

Here are examples of sneaky budget-busters:

- Skimping on the foundation (building a house on Jell-O isn't a great idea)
- Buying cheap fasteners, hinges, latches (rust and breakage are not your friends)
- Forgetting about ventilation (plants love fresh air as much as we do)

Think of your greenhouse as an investment in your garden's future. Like any good investment, it's about making smart choices upfront that pay dividends in homegrown tomatoes later.

That said, there's plenty of room for ingenuity and balance. Some of the most productive greenhouses I've seen were built with more creativity than cash.

In general, mini versions (cold frames, mini-greenhouses) are great for experimenting with upcycled materials and budget-friendly hacks.

For bigger projects, overall you want structural and mechanical stuff to be good quality, while things that fill the insides can be more creative. Think: could I replace it easily if my DIY upcycling idea went sideways?

Here's a basic guide to bargain hunting in three categories, like a "traffic light" for decisions: green light for creative reuse, yellow for careful bargain hunting, and red for "don't mess around with this stuff."

1. **Green: Great for upcycling projects**: Shelving and benches, rainwater collection, planters and pots, pathway materials, and decorative elements. Also creative insulation solutions like bubble wrap, old carpets, and cardboard.
2. **Yellow: Quality at a bargain**: Frame materials, covering, basic insulation, simple ventilation fans, doors and windows, and starter tools. Note: still make sure these are in excellent condition.
3. **Red: Worth investing in quality:** Foundation materials, seals and weatherproofing, ventilation systems (beyond basic fans), fasteners & connectors, and equipment (heating & cooling, irrigation).

By knowing where to save and where to splurge, you can build a greenhouse that's budget-friendly without sacrificing durability or performance.

Final Thoughts

Hey, I know the feeling when you walk into a garden center and your wallet starts sweating. Believe me, I get it. Maybe we can ease that anxiety.

Think of building your dream greenhouse like making a sourdough starter—you don't create a bakery on day one. Start with something simple, like a cold frame or mini greenhouse, which lets you test growing techniques without major commitment.

As your confidence (and harvest) grows, reinvest your grocery savings into upgrades. Maybe you can build a basic lean-to one season, add automation the next season, and expand your growing space the following year.

At the time of writing, a basic 8x8 greenhouse can cost anywhere from $500 to $3,000 or more, depending on materials, features, and labor costs. Spreading these investments over 2–3 years makes it more digestible. Plus, you're learning what features you actually need rather than paying upfront for bells and whistles you might never use.

Remember: many a great gardener started with nothing more than a simple cold frame and a dream. Your greenhouse empire can grow just like your plants—one step at a time.

Hopefully you're now armed with budget-friendly strategies to make a shopping list that won't terrify your credit card. But before you start collecting materials or mapping out your DIY masterpiece, let's talk about one more crucial piece: where you're going to put it.

Maybe you can envision your greenhouse already, or maybe you're still staring at the selections, wondering where the heck to start. Either way, picking a location will help you! Understanding your site's specifics narrows down your greenhouse options, and it helps you have clearer pictures as you daydream.

CHOOSING THE RIGHT LOCATION: SUNLIGHT, SPACE & STRUCTURE

Time to find a good spot for your greenhouse. It doesn't have to be perfect, but there are some things to look for. You're like a real estate agent for plants, and your clients have some boxes to check. They want plenty of natural light, good drainage, and easy access to utilities. Plus, they're planning to stay for the long haul, so this better be good.

Sun

First up: sunlight, the ultimate currency in greenhouse real estate. Aim for an area that offers **full sun**, which means at least six hours of direct sunlight per day.

Some plants appreciate a bit of shade, so consider the mix of your intended greenery. The beauty of a greenhouse is that you can create different areas inside of the greenhouse that get more or less light.

Here's what most guides won't tell you—not all sunny spots are created equal. That patch that's gloriously bright in summer might turn into a shadow realm come winter, thanks to the sun's seasonal pattern shifts.

One solution is to track your sun patterns through the seasons. Sounds tedious, I know, but it's less work than relocating an entire greenhouse. Mark potential spots and check them at different times of day.

Or if you don't have that long to wait, use a sun-tracking app or resources online. Check out https://www.sun-direction.com/ for a cool interactive map that shows you sun direction and day lengths in your exact location throughout different times of the year.

Access

Now let's talk about access, because enthusiasm for hauling supplies tends to fade fast.

Your greenhouse needs three key connections: water, electricity (if you're planning on heating or ventilation), and you. That last one might seem obvious, but I've seen enough awkwardly placed greenhouses to know it needs saying. Position your greenhouse where you'll actually visit it, even in less-than-perfect weather. If accessing your greenhouse requires an expedition worthy of National Geographic, you might want to reconsider the location.

Also, put your greenhouse within easy reach of a water source. I once thought, "Oh, carrying water isn't that bad!" Narrator: It was that bad.

Make sure there's a clear path, free from obstacles, to navigate to and from your greenhouse easily, especially when your hands are full of gardening tools or that first glorious harvest.

Inside, accessibility becomes even more critical. You'll want enough room to move around without feeling like you're in a game of gardening Tetris. We'll go over interior design even more in the next chapter.

Foundation

Here's something people often forget: the ground needs to be level. This is necessary to ensure structural stability, efficient drainage, even temperature control, longevity, and ease of use.

However, greenhouses are all about adapting. If you are on sloped land and are up for the investment, fear not! There's a solution. An **uneven-span** greenhouse works great here. Like a lopsided peaked roof, one side of the slope is longer than the other.

This design isn't just for looks, either. The longer slope typically faces south to maximize sunlight exposure, while the shorter northern side helps minimize heat loss. (This is for the northern hemisphere. Reverse that direction for southern hemisphere applications.)

Wind

Wind exposure is another factor. While a gentle breeze is a greenhouse's best friend, strong gusts can be its worst enemy. Choose a location with some natural protection, like a hedge or fence, especially if you live in a particularly windy area. This extra shield can provide peace of mind when the weather turns blustery.

Zoning and Permits

Now, before you break ground, there's the not-so-glamorous side of greenhouse planning: permits and zoning. Your local ordinances might just have a word or two about where and how you construct your greenhouse. It's wise to check your local zoning regulations to ensure you're in the clear. This might involve a quick call to your town office or a perusal of their website. Permits might be necessary, especially for larger structures, so it's best to tackle this paperwork early to avoid any headaches down the line.

Finding Your Spot in the Sun

Choosing your greenhouse location is like picking the perfect spot at a coffee shop—you want that sweet spot with just enough sun, away from the chaos, and close enough to the amenities (in this case, your water source).

In choosing your location, you're laying the groundwork for everything that follows. It's more than just finding a sunny spot; it's about creating an environment where your plants—and you—can thrive. The right location sets the stage for a successful greenhouse experience, turning that initial vision into a vibrant reality.

CHECK-IN

Wow! You've toured the greenhouse showroom in your mind and explored material options (maybe more than you knew existed). Look at you go! By working through this chapter, you've accomplished a lot already! You can now:

- Identify the major greenhouse styles and match their strengths to your specific climate challenges

- Evaluate covering materials based on practical factors like durability, light transmission, and insulation value
- Understand the pros and cons of different structural materials, from affordable PVC to long-lasting galvanized steel
- Recognize the importance of proper site selection, evaluating things like sun patterns and accessibility
- Make budget decisions from an informed perspective, knowing where to save and where to invest
- Think about your greenhouse as a long-term, potentially expandable project

Even if you haven't finalized every decision, you're now equipped with a solid framework for making informed choices rather than stabbing in the dark. That's huge progress from where you started!

Let's see where your greenhouse blueprint stands:

- Have you narrowed down your greenhouse type to 2–3 realistic contenders? (So long, fantasy crystal palace with built-in espresso bar ... at least for now ...)
- Do you have a sense of which covering materials (Glass, polycarbonate, etc.) match your climate challenges, plus other needs and wants?
- Have you considered where your greenhouse will live in your outdoor space?
- Have you thought about DIY versus prefab kit options that align with your skills and time? (Both are good—just find the one that works for you.)

Not hitting all these marks yet? Don't sweat it! Make note of your initial ideas, and know that they'll develop.

In the next chapter, we'll help you narrow things down further as we talk about setting up your space. The perfect frame and covering materials will become clearer once you understand your interior layout needs.

Before we move on, here's a helpful action you can take. Take a moment to write down basic budget and materials list as it stands now. Future you will be grateful—trust me!

Here's how to start. Write down:

- Your "must-have" features
- Your "nice-to-have" features
- Your absolute maximum budget
- Your comfortable spending range
- Any materials you know you'll need

How do these align? If there's a gap, no worries—we'll be exploring creative solutions in the next chapter, and throughout the book, and you will encounter plenty of options in your greenhouse journey.

Quick success tip: Once you know what and where you're building, and what materials you'll need, you can start gathering materials before you need them. It's like meal prepping, but for construction. Your wallet (and stress levels) will thank you.

Wrap-Up

As you're sketching out plans and making your shopping list, remember something important: whether you're working with salvaged windows or top-of-the-line polycarbonate, the heart of greenhouse gardening isn't in the materials. It's in the memories you'll make, the passion you'll pour into the project, and the victories you'll celebrate in your very own sunshine sanctuary.

There's a unique joy in knowing that every beam you raise and nail you hammer isn't just building a greenhouse—it's building your self-sufficiency, sense of accomplishment and contribution, and your own little rebellion against winter, one satisfying *thunk* at a time.

So make your choices and enjoy the thunking.

Next up: let's turn that empty greenhouse into your dream garden! First, we'll deck out the interior with essential tools (and some fun extras), then tackle the nitty-gritty of climate control and construction.

Let's continue fleshing out the dream!

3

INSIDE THE GREENHOUSE

HOW TO DESIGN, ORGANIZE, AND EQUIP YOUR GROW SPACE

Remember that feeling of moving into your first apartment? Standing there in an empty room, pizza box in hand, wondering how you're going to transform this blank space into something that doesn't scream "just escaped my parents' basement"?

Well, congratulations! You're about to have that same exciting-yet-mildly-terrifying moment with your greenhouse—minus the questionable takeout. (Or keep the takeout, whatever inspires your art!)

Now that we've imagined the exterior and location of your greenhouse, let's take a moment to imagine what will go *inside*.

DESIGN YOUR DREAM GREENHOUSE: LAYOUT & AESTHETICS

The secret to a successful greenhouse isn't cramming in every plant from your wishlist (though the temptation is real). It's about creating an organized space where both you and your plants can thrive.

The Zones of Your Greenhouse

Creating a functional layout is crucial for maximizing space and nurturing plant growth. Think of your greenhouse like a well-organized kitchen, where zonal planting areas serve as different cooking stations.

For example, it's a great idea to have a prep station for sowing seeds and potting plants. And don't forget storage—a place for tools and supplies ensures you're not rummaging around like a squirrel looking for its last acorn.

Like a master chef, you'll learn your space the more you go, too, finding the perfect spots for each plant's needs. It's a pretty cool feeling as you get to know your zone. See that sunny section over there? Perfect for tomatoes and peppers to bask. While shade-loving ferns and salad greens find solace in that cooler corner.

As you start to envision your space, here are things you'll want to consider including:

1. **Growing zones**: Designate specific areas based on plant needs:
   - Sunny spots for sun-loving plants
   - Shadier areas for plants that prefer less direct light
   - Vertical spaces for climbing plants
   - Ground-level beds for root vegetables
   - Think easier access for higher maintenance or more frequently harvested plants.
2. **Work area**: Create a dedicated potting station where you can:
   - Store your essential tools within arm's reach
   - Have a comfortable surface for transplanting and seeding
   - Keep soil and amendments easily accessible
3. **Storage solutions**: Implement smart storage to keep everything organized:
   - Wall-mounted tool racks
   - Waterproof containers for soil and amendments
   - Shelving for pots, trays, and supplies

Here's a basic summary: consider setting up dedicated areas for propagation, growing, and storage.

Workflow and Accessibility

Designing for workflow and accessibility is key. Think of pathways like the highways of your greenhouse—you want them wide enough for both your wheelbarrow *and* your victory dance (definitely important). Also, consider any accessibility you might need, like for a wheelchair or any equipment you plan on using.

Like that master chef's kitchen, you want everything within arm's reach. Your back will thank you for raising those workbenches to an ergonomic height where you're not hunching over like a wilted bean sprout every time you pot a plant. Remember: a comfortable gardener is a happy gardener, and happy gardeners grow happy plants. It's basically science.

If you are thoughtful about your design, it allows you to glide smoothly through your tasks, making gardening a pleasure rather than a chore.

Maximizing Your Space

Even if your greenhouse isn't quite as spacious as Buckingham Palace, you can make every square inch count.

Here are some ideas for creative usage of space:

- **Vertical space**: Install adjustable shelving and hanging systems
- **Corner spaces**: Use corner shelves or triangular benches
- **Under-bench storage**: Create storage solutions beneath work surfaces
- **Door space**: Consider over-door shelving or hanging baskets

Flexibility and adaptability are your allies in designing zones that can evolve with your needs. Modular shelving or tables offer the freedom to reconfigure your setup as your plant collection grows or changes. Movable planters can be repositioned with the seasons, adapting to the changing light and temperature conditions.

Light Management

Optimizing light distribution is as essential as ensuring everyone at the dinner table gets a slice of pie. Here are some ways to optimize light:

- Position taller plants along the north wall to prevent shadowing (it's like putting the tall kids in back for the class picture, only instead of a camera it's the sun). (Reverse this if you're in the Southern Hemisphere.)

- Use reflective surfaces on walls to bounce light to darker corners
- Install adjustable shelving to modify height based on plant needs
- Consider supplemental lighting for winter months or cloudy climates (we'll talk more on this later).

Quick success tip: Photograph your greenhouse interior (or where you imagine it will be) in the morning, midday, and late afternoon to track sunlight patterns before finalizing your layout. These simple snapshots will save you from discovering too late that your seedling station sits in afternoon shade.

You're the Artist!

But a greenhouse isn't just about functionality—it's your sanctuary! Let your inner Monet loose by adding pops of color and texture. Decorative plant arrangements can create focal points, drawing the eye and adding character. Arrange plants in clusters for dramatic effect, toss in a statement pot or two, and maybe even hang a wind chime.

Let your personality shine through with unique touches that make the space truly yours. Because nothing says, "Welcome to my greenhouse," like an inviting space that's as delightful for you as it is for your leafy guests.

Want to play architect before committing to that trellis placement? Sketch your ideas on paper, drawing out where different zones could be. You can use graph paper for easy, consistent measurement.

If you're feeling fancy, try one of those 3D garden planning apps—they're like The Sims for plant people. Software for 3D modeling can offer great tools to design and give you a virtual tour of your planned space.

As you plan, remember that this is *your* place to grow plants. And also to cultivate confidence, joy, creativity, and a deep connection with nature. So make it yours!

MASTER TOOL GUIDE

Let's talk about the tools you'll actually use—not the fancy gadgets that'll end up collecting dust like a neglected windowsill.

There are some must-have basics, beginning with hand tools. Pruners are essential for keeping those overzealous stems in check, while trowels help you nestle seeds snugly into the soil. They're like the salt and pepper of gardening—basic, necessary.

And as any seasoned gardener will tell you, a good pair of gloves is worth its weight in gold, keeping your hands safe and splinter-free.

Let's break down some of the essentials. I'll throw a couple fancy gadgets in there, too. For the most part, though, the fancier stuff comes later.

1. Foundational Hand Tools

These are your daily companions in the greenhouse—the tools you'll reach for most often:

- **Hand trowel & transplanting spade**: For planting, digging, and moving soil.
- **Pruning shears**: Essential for trimming plants and managing growth.
- **Garden fork**: A pitchfork's smaller cousin. Useful for loosening soil and mixing compost.
- **Spade**: For heavier digging and moving soil efficiently.
- **Garden knife**: Versatile for cutting twine, slicing through tough stems, or even dividing root-bound plants.
- **Heavy-duty gloves**: To protect hands during planting and maintenance.

2. Watering
(More details in Chapter 5) ...

While the "dump and hope" school of watering certainly has its merit, equipping yourself with the right watering tools can make all the difference in meeting your plants' hydration needs, as well as making your life a lot easier.

We'll cover the essential tools here, and in Chapter 5, we'll dive into more advanced irrigation systems and automated watering solutions that can really level up your watering game.

The Essentials

1. **Watering can**
   - **Why you need it:** A simple and versatile tool for targeted watering, ideal for small areas or delicate plants like seedlings.
   - **Features to look for:**
     - A fine rose attachment for a gentle spray.
     - Lightweight but sturdy construction.
2. **Garden hose with adjustable nozzle**
   - **Why you need it:** For larger greenhouses, a hose allows you to water multiple plants efficiently.
   - **Features to look for:**
     - Adjustable spray patterns (e.g., mist, shower, jet) for different needs.
     - A long hose to reach all corners of the greenhouse.
     - Maybe a hose reel or holder to keep the hoses organized.
3. **Soil moisture meter**
   - **Why you need it:** To avoid overwatering or underwatering by gauging soil moisture levels accurately.
   - **Features to look for:**
     - Easy-to-read moisture level indicator.
     - Models with added features like light and pH measurement for comprehensive soil health monitoring.
4. **Misting system *(fancy)* or spray bottle *(not fancy)***
   - **Why you need it:** To increase humidity levels and provide gentle watering for tropical plants, seedlings, or delicate foliage.
   - **Best for:** Greenhouses growing orchids, ferns, or other moisture-loving plants.

Optional But Useful Watering Tools

1. **Wicking mats**
   - **Why you need it:** These mats keep seed trays and pots consistently moist by drawing water from a reservoir below.
   - **Best for:** Maintaining even moisture for seedlings or shallow-rooted plants.
2. **Handheld wand or watering lance**
   - **Why you need it:** Ideal for reaching hanging plants or plants at the back of your greenhouse without straining.

- **Features to look for:**
    - A long, flexible arm.
    - Adjustable spray patterns.
3. **Self-watering pots and trays**
    - **Why you need it:** This is an awesome little invention that you can buy pre-made or make yourself with a few DIY videos and simple supplies. Just look for "self-watering" or "self-wicking" containers. These containers have reservoirs that provide a steady water supply to plants, reducing manual watering frequency.
    - **Features to look for:**
        - Built-in reservoirs.
        - Easy-to-monitor water levels.

3. Maintenance and Safety

Your poor man's greenhouse insurance policy—they keep both you and your plants protected:

- **Cleaning supplies**: Brushes and eco-friendly cleaners to maintain a hygienic greenhouse.
- **Goggles**: For handling chemicals or certain maintenance tasks.
- **Repair kits**: Duct tape, zip ties, and basic tools (hammer, multi-head screwdriver, sealant, etc) for quick fixes.
- **First aid kit**

4. Organizational Tools

Because a tidy greenhouse is a happy greenhouse.

- **Shelving or tables**: For keeping plants organized and maximizing space.
- **Labels and markers**: To identify plants and track planting dates.
- **Storage containers**: For tools, seeds, and fertilizers.
- **Buckets or trays**: Multipurpose containers for carrying tools, collecting weeds, or soaking seeds before planting.

5. Precision and Specialty Tools

These tools help you perform specific tasks with greater accuracy:

- **Hori hori knife**: A versatile tool for digging, cutting, and weeding, with a built-in measuring guide for planting depth.
- **Weeder**: For removing weeds with precision, especially those with deep roots.
- **Dibber**: Handy for making uniform planting holes for seeds and bulbs.

6. Harvesting Tools
(More details in Chapter 9)

Your toolkit for gathering the fruits of your labor:

- **Sharp scissors or garden snips:** For precise cuts on herbs and tender stems
- **Pruning shears:** For woody stems and thicker growth
- **Garden knife (hori hori):** For root vegetables and tough stems
- **Collection baskets:** For safely gathering and transporting produce
- **Clean gloves:** For protecting both you and your harvest

7. Optional But Handy Tools

These tools aren't essential, but they can make your greenhouse life easier:

- **Propagation mats**: To provide bottom heat for seed starting.
- **Transplanting tongs**: If you do a lot of seed starting and seedling transplanting, these make it really easy.

Expanding Your Arsenal

There are all kinds of variations of full-size and hand gardening tools, from three-prong cultivators and hand rakes to trowels, scoops, and spades. Not sure which ones you need? Just strike up a conversation with any gardener about their favorite tool. Fair warning: we're like kids showing off our favorite toys at recess.

Coming up, we'll dive into all sorts of fun tech on the more "advanced" tool list, like climate control tools such as thermometers and hygrometers. These tools keep a vigilant eye on temperature and humidity, ensuring your plants remain in their comfort zone.

In the coming chapters, we'll also discuss lighting tools, automatic irrigation and watering systems, pest management tools, soil and fertilization, and other gadgets to make your growing life easier.

Finding quality tools and equipment doesn't have to mean shelling out big bucks. Local gardening centers often offer a range of products to fit every budget. Garden centers are also like treasure hunts where the staff actually want to help you find the gold. Plus, they love sharing local growing tips like gardening cheat codes.

Online shopping works, too. Remember to read those reviews, and go with something mostly good. It's kind of entertaining reading a passionate gardener's one-star review of a flimsy trowel.

Tool TLC 101

Think of your tools like a chef's knives, which need to stay sharp to slice through tomatoes instead of squishing them. Similarly, your garden tools need love to keep performing their best, and a little maintenance goes a long way. A quick clean after use (yes, every use—I learned this the hard way), occasional sharpening, and a bit of oil to keep rust at bay will keep them happy.

Store them somewhere dry and organized, like a pegboard wall of fame, and they'll be ready for action whenever inspiration strikes.

Your Tool-Care Cheat Sheet:

- Wipe 'em down (dirty tools are sad tools)
- Keep 'em sharp (dull blades are like using a spoon to cut bread)
- Oil the metal bits (rust is not vintage patina)
- Store 'em proper (not in that pile behind the greenhouse—again, learned the hard way)

INSIDE THE GREENHOUSE

WHERE FUNCTION MEETS BEAUTY

Designing your greenhouse isn't just about squeezing in as many pots as possible—it's about creating a space that works *and* wows. A space that flows with your daily rhythm, but also lifts your spirits when you step inside.

When form and function come together, you get a greenhouse that not only grows food but feeds your soul. Let's talk about a few creative strategies to maximize both sides of your greenhouse design.

Maximizing Space with Vertical Gardening

Remember that old saying about the only way to go is up? Well, in greenhouse gardening, you reach a point where that's true. Think of your greenhouse as a multi-story plant hotel, where every level offers a different view and opportunity for growth. Vertical gardening is like giving your plants a penthouse suite.

By stacking your greenery, you effectively multiply your growing space, which is particularly handy if your greenhouse is on the petite side. Imagine vertical planters and shelving units where you cultivate strawberries cascading down like nature's jewelry. Or a wall of herbs arranged in levels like the world's tastiest theater seating. These structures maximize space while also adding a dynamic visual element to your greenhouse, turning it into a living work of art.

Trellising is another nifty trick, perfect for climbing plants like tomatoes and cucumbers. The vines stretch upwards, reaching for the sky while saving precious floor space for other leafy companions.

The beauty of vertical gardening lies in its ability to produce a higher yield per square foot, making it a favorite for those working with limited ground area. It's like squeezing the last bit of toothpaste from the tube.

For the creatively inclined, innovative solutions abound. Consider crafting a pallet garden, a DIY project that upcycles old wooden pallets into a lush vertical planter. Or turn old gutters into skyrise salad bars.

For those seeking a more high-tech approach, hydroponic towers offer a compact and efficient way to grow plants without soil. These towers circulate nutrient-rich water, providing everything your plants need to thrive in a tidy, space-saving column.

Vertical gardening isn't just about cramming more plants into a small area; it's about finding possibilities and joy as we reimagine how we use space.

Choosing the right plants is key to vertical gardening success. Vining vegetables such as beans and peas naturally lend themselves to this method, happily climbing their way to the top. Herbs like basil and mint thrive in vertical arrangements, offering easy access for a quick snip when you need quick ingredients.

Don't worry, we'll delve into more plant choices later. Just remember: there are possibilities. When you think you've run out of space, just look up.

Incorporating Aesthetic Elements in Design

Let's talk greenhouse glamour! Sure, functionality is key, but who says your plant paradise can't also be worthy of a magazine spread? (Or at least your social media feed.) Consider your greenhouse as more than a space for growing plants—it's an expression of your personality.

Start with pathways that don't just get you from A to B, but make you feel like you're strolling through a botanical garden. Wood chips or stepping stones aren't just practical; they're like nature's red carpet leading you to your next gardening adventure.

And speaking of nature's palette—mix those plants up like you're painting with chlorophyll! Pair those deep purple basil leaves with bright green lettuce, and suddenly you're not just growing lunch, you're creating art. Coordinating plant colors with greenhouse interiors can create a cohesive and harmonious look. Bright blues and yellows can evoke a sunny day, while cool greens and whites offer a calming vibe.

Now for the fun part: personality! Personal touches and themes are what transform a greenhouse from a mere structure into a sanctuary. Maybe that means displaying Grandma's vintage watering can, or that hilariously ugly garden gnome you can't bear to part with (be proud of it). Cultural touches? Absolutely! Plant that curry leaf tree next to your Italian herbs—it's like hosting an international plant party.

Whether you like modern or rustic, or whether you're a "everything in its place" minimalist or more of a "controlled chaos is still control" naturalist, your greenhouse should reflect what makes your gardening soul sing.

Textured pots and planters add layers of interest, drawing attention and inviting touch. Maybe your greenhouse vibe involves running your hands over a terracotta pot's rough surface or the smooth curve of a glazed ceramic. These elements engage the senses, making your greenhouse a visual feast and a tactile one too.

And don't forget to engage your nose in the design. Imagine that, a greenhouse that engages all five senses! Tuck some fragrant flowers like lavender or jasmine near your seating area. It's like nature's aromatherapy, except instead of buying expensive oils, you're just letting your plants do what they do best (making the bees swoon).

Create yourself a little retreat spot too. I'm talking about a cozy corner where you can sip your morning coffee while admiring your tomatoes' progress. Create your own luxury box seats by adding cozy seating—because hey, you deserve a front-row view of the botanical show you're directing.

Add a small fountain if you're feeling fancy—nothing calms the soul quite like the sound of trickling water mixing with the rustle of leaves. Though fair warning: you might find yourself inventing excuses to "check on the plants" just to spend more time in your slice of paradise.

Remember, this isn't just a growing space—it's your happy place. Make it somewhere that brings a smile to your face, even when that experimental pepper plant is giving you attitude. Because let's be honest: a greenhouse should grow joy just as much as it grows vegetables.

SETTING A REALISTIC BUDGET: COST-SAVING TIPS & TRICKS

Let's chat again about everyone's favorite topic: money! (I heard that groan.) It's time to bring up your budget spreadsheet, and add the new things you've learned in this chapter.

Think of your budget like a road trip. You need to have gas money (the essentials), snack money (the fun stuff), and a little set aside for when that weird dashboard light starts blinking (unexpected surprises). Your budget guides you, keeps you on track, and occasionally lets you take a detour for that irresistible garden gnome.

As you continue to add to this budget, you'll also need to start distinguishing start-up costs from ongoing expenses. Start-up costs include materials, tools, and any professional help you might need. Ongoing expenses, like water and electricity, ensure your greenhouse remains a thriving haven.

Always pad your budget with a little extra because, trust me, Murphy's Law loves gardening almost as much as we do.

Let's do some more treasure-hunting! Second-hand equipment can be garden gold. Check online marketplaces, garage sales, or that neighbor who impulse-bought a greenhouse and decided growing cacti was more their speed.

And timing is everything—shopping seasonal sales is like extreme couponing for gardeners. That premium potting bench might just become affordable when stores are making room for snow blowers.

A little resourcefulness goes a long way when adapting household items for greenhouse use. Old milk jugs can become watering cans, and that ancient, single-glove sock can keep your hand warm while planting. An old ladder can serve as a trellis. Bada-bing.

Don't forget, you can even get creative with funding! I'll mention it again: some communities offer grants for sustainable projects (yes, really!). It's like applying for a scholarship, except instead of writing essays about your life goals, you're planning to grow organic tomatoes.

Low-interest loans for home improvements can also be a viable path, spreading the cost over time instead of an upfront hit to your wallet.

I know a well-planned budget can feel restricting, but it can actually help you achieve your goals. This isn't about pinching pennies until they squeak; it's about making smart choices so you can splurge on that automated watering system that'll let you actually go on vacation someday.

Remember: building a greenhouse is an investment in your future salads (and sanity). With a clear financial plan, you're well on your way to turning your vision into reality.

Check-In

You've graduated from dreaming of your greenhouse from the outside, to actually going in and planning its inner workings—that's huge progress! Look what you've accomplished:

- You can now design functional zones that maximize your growing space.
- You understand how to make things accessible, and the principles of ergonomic workspace design that will save your back and make daily tasks enjoyable.
- You understand how light patterns affect plant placement within your greenhouse.
- You've learned which essential tools will make your greenhouse life easier (without cluttering up your space), and how to take care of them.
- You can identify vertical growing opportunities that multiply your planting area.
- You've discovered how to incorporate aesthetic elements that make your greenhouse not just functional, but enjoyable.

... and much more.

These skills transform your greenhouse from an empty structure to a well-organized growing machine that's also a joy to spend time in.

Let's see where your planning stands:

- Have you sketched a basic layout showing different functional zones? Growing areas, work space, storage? (Remember to include walkways wide enough for both you and your tools.)
- Have you created a prioritized tool list distinguishing between immediate needs and future acquisitions? (Start with the foundational items.)
- Have you located your water source? (Trust me, this matters more than you think.)
- Have you imagined at least one vertical growing opportunity, or aesthetic addition, you might like to add to your design?
- Have you made note of at least one money-saving strategy that resonates with you?

Not quite there on some points? No problem! Take a moment to sketch a rough draft of your layout if you haven't already. Don't worry, this isn't final. You can toss it right after you do it, and try again.

Even basic planning or experimenting at this stage will pay dividends later, and your blueprint will naturally evolve as you move forward.

Quick success tip: When sketching your layout, you can use tracing paper overlays to experiment with different arrangements without starting from scratch each time. Keep the best versions—they'll be valuable reference guides during construction.

Wrap-Up

Now that we've got your greenhouse selected, organized, and basically equipped, you might be wondering how to keep all those lovely future plants happy. The secret lies in two crucial factors: climate control and proper watering.

In the next chapter, we'll dive into the art of climate control—because your greenhouse shouldn't feel like a sauna in summer or an ice box in winter. We'll explore everything from basic ventilation to advanced temperature management techniques that'll keep your plants living their best lives, no matter what nature throws your way.

4

CLIMATE CONTROL

CREATING THE IDEAL ENVIRONMENT

This, my friend, is the real peek behind the curtain of greenhouse gardening. Climate control is where your greenhouse truly transforms from a simple structure into a thriving, self-contained ecosystem.

It's like running your own botanical Broadway blockbuster—where your plants are the stars of the show. On our stage, you control the lights and the wind. Temperature sets the mood, and humidity shapes the atmosphere. Just like a director crafts the perfect ambiance to create a scene that immerses you in a new world, you'll be orchestrating these elements to create show-stopping performances from your plants. Thankfully, these performers work for water and sunlight instead of demanding their own dressing rooms. Although, I'll warn you, orchids and gardenias are notorious divas ...

To get that standing ovation, there are a few important factors to manage. While managing temperature gets a lot of attention — and for good reason — it's only part of the story. To create the perfect environment for your plants, you'll also need to balance humidity, airflow, light, and water.

In this chapter, we'll break down each of these essential factors, empowering you to turn your greenhouse into a haven where your plants flourish year-round. If you're new to all these concepts, take it one step at a time.

I've definitely wilted plenty of plants on the climate control learning curve. But I learned, and as my plants revived, so did my confidence. And I keep getting better with every season (at least that's what I tell myself when I look at the white hair spreading across my beard).

So take a breath, smile and remember, I've got your back, and you've got this! You've already learned a lot you didn't know before, right? It's basically second nature to you.

Ready? Let's dive in!

GREENHOUSE MONITORING TOOLS: YOUR EARLY WARNING SYSTEM

Your plants can't text you when they're too hot or cold. That's where monitoring systems come in. They're your eyes and ears when you can't be there, helping prevent those "oh no" moments when you return to discover your greenhouse is a sauna or an ice box.

You'll want to start with the basics: digital thermostats and hygrometers. These devices measure temperature and humidity levels, ensuring both remain within the cozy confines plants love.

Let's look at some more devices that make greenhouse living easier.

Essential Temperature Monitoring Tools

- **Digital thermometers** with high/low temperature limits let you check the greenhouse's temperature, and if connected to a temperature control system, they will automatically make sure things stay in a healthy range. Some models even send you updates on your phone.
- **Remote sensors** monitor temperature in specific areas of your greenhouse so no area becomes too hot or cold. They send real-time data to the main thermostat, maintaining even and plant-friendly conditions while saving energy. The thermostat can average the temperatures from all the sensors and adjust the temperature to keep the whole greenhouse comfortable.
- **Infrared thermometers** are another convenient tool for spot-checking surface temperatures around your greenhouse to make sure no corner is too hot or too cold.

Set-Up Tips

Setting up these monitoring systems is as crucial as choosing the right ones. Calibration is your first step, ensuring the readings are accurate. Think of it as tuning a guitar; just a few tweaks, and everything sounds—or in this case, reads—perfect.

Place your sensors strategically throughout the greenhouse to gather representative data. You wouldn't put all your thermometers in the sunniest spot and expect an accurate overview, right? Spread them out so you can obtain a general idea of your greenhouse's microclimate, just as a photographer will take different photographs from different vantage points for the best shot.

Here are some set-up specifics:

1. Place sensors at plant level, not ceiling height (that's where your plants live, after all).
2. Definitely install multiple sensors if your greenhouse is larger than 100 square feet.
3. Keep sensors out of direct sunlight for accurate readings.
4. Calibrate all devices before first use, ensuring readings are accurate. Think of it like tuning a guitar before playing it.
5. Keep extra batteries and add a battery check to your monthly maintenance schedule.

Maintenance Know-How

Connectivity problems with smart devices can definitely happen, so keep an eye on connectivity. You know the classic "my smart device isn't so smart" scenario. It's like having a teenager who suddenly stops responding to texts right when you need them most. In this case, it's a monitoring device ghosting you during a cold snap. Let's avoid that drama in your greenhouse, shall we?

If your device isn't connecting, a quick restart of the router or checking the Wi-Fi signal might do the trick.

Keep a stash of fresh batteries tucked away like emergency chocolate (except these won't melt when you need them most). Set yourself a monthly reminder to check those battery levels. After all, discovering dead batteries during a cold snap is about as fun as finding out you're out of coffee on a Monday morning!

Troubleshooting Tips Summary

- Wireless sensors acting up? → Check battery levels first.
- Readings seem off? → Recalibrate against a known accurate device.
- Smart device not connecting? → Reset your router or check signal strength.
- Different sensors showing varying readings? → Map cold/hot spots in your greenhouse with infrared. These can be used strategically as different "microclimates" or measures can be taken to balance out temperatures.

Humidity Monitoring Tools

A **hygrometer** is basically a humidity detective for your greenhouse. Think of it as a weather station that only cares about moisture in the air—like that one friend who's obsessed with having the perfect hair day. (Hey, we all have our things.)

Nowadays, most digital hygrometers have a display showing the relative humidity percentage, and many also include a temperature reading as a bonus feature.

Setting one up is surprisingly simple:

1. Place it at plant height, away from direct sunlight or water sprays (it wants to measure the air, not take a shower)
2. Let it adjust for about 2 hours before trusting the reading
3. For accuracy's sake, calibrate it using the salt test: put it in a sealed container with a small dish of wet salt for 8–12 hours. It should read 75% humidity. If it doesn't, adjust according to your model's instructions (or just note the difference)

Want to get fancy? Grab one with a remote sensor and phone alerts. Then you can obsess about your greenhouse's humidity levels from the comfort of your couch.

And let me tell you—nothing says "I've leveled up as a gardener" quite like texting your friends about your perfect humidity readings instead of your latest Netflix binge! (No judgment; we all have our things.)

Pro tip: Different plants like different humidity levels. If you're growing a mix of plants (and who isn't?), you'll be aiming for a sweet spot around 60–65%. More on that later.

Final Thoughts

Monitoring systems may seem like a high-tech luxury, but they're worth every penny for the peace of mind they provide. They allow you to step away from your greenhouse without worrying about your plants, because these tools do the worrying for you.

Trust me, the moment you're snuggled up on your couch during a storm, checking your greenhouse temps from your phone instead of making a mad dash through the rain? That's when you'll be sending me mental thank-you notes.

With these systems in place, your plants will reward you with steady, reliable growth: lush, low-drama, and full of life.

REGIONAL CONSIDERATIONS: A QUICK NOTE

While the core principles of greenhouse gardening are universal, your local climate adds its own special quirks to the equation. It's kind of like how everyone needs shoes, but a surfer needs different footwear than a mountain climber. Let's look at some examples to get a clearer picture of how your particular region can affect choices.

Desert dwellers in Arizona might need to focus on triple-layer shade cloth and misting systems, while Minnesotans should prioritize heavy-duty insulation and snow-load reinforcement. In the Pacific Northwest, managing humidity and ventilation becomes crucial to prevent fungal issues, whereas gardeners in Texas might need to incorporate geothermal cooling techniques like underground air pipes. Coastal gardeners battle salt spray and wind, while inland areas contend with temperature extremes.

Check with your local extension office or garden club for specific regional recommendations. If you're in a humid area, consider adding about 25%

more ventilation than you think you need. In regions with high-temperature swings, consider increasing your thermal mass by 30%.

Jot down the unique challenges you hope to address with your greenhouse and rank them on how pressing it is to solve them (what are your region's biggest quirks for gardens?). This can become your roadmap for prioritizing greenhouse investments.

TEMPERATURE CONTROL: THE GOLDILOCKS ZONE

Your plants aren't that different from us—they like their environment "just right." Consistent temperatures ensure they grow strong and healthy, while wild swings can stress them out, affecting their metabolism, hindering growth, and making them more susceptible to disease.

Think of it like you trying to sleep through a night of constant temperature changes. You'd wake up grumpy, and so do plants.

Different plants have different ideal temperature ranges, but most common greenhouse varieties thrive between 65°F and 75°F (18°C and 24°C).

Now, if you're thinking, "Boy JR, this sure seems like a lot to manage," think back to when you learned to drive a car. Initially, it feels like there are SO many things you need to watch—the speed, the mirrors, the other vehicles. But after a while it becomes second nature, and you're doing all the adjustments without even thinking, right?

That's exactly how it ends up being with managing your greenhouse's climate. Sure, you'll have some bumps along the road (ask my wilted plants), but every little "oops" makes you more skilled at reading your greenhouse's cues.

Temperature control in your greenhouse is a well-choreographed dance between passive methods (like thermal mass and strategic ventilation) and active systems (such as fans and heaters). The key is letting passive methods do the heavy lifting while active systems step in for backup.

Just as you wouldn't blast your AC with the windows open, start with smart passive design choices like proper orientation, thermal mass placement, and good insulation. Then, layer in active controls to fine-tune when nature throws a curveball. By combining both approaches strategically, you'll create a more resilient and energy-efficient growing environment that won't have your electricity bill doing backflips.

Passive Temperature Control

Think of these methods as your first line of defense—they're like having a smart thermostat that doesn't need batteries. The trick is combining them strategically based on your climate and greenhouse design.

Thermal mass, for instance, is a secret weapon in the temperature control arsenal. Think of thermal mass like your greenhouse's built-in heat bank. Things like water-filled barrels, masonry bricks, concrete blocks, or even a stone walkway can soak up warmth when the sun's out and gently radiate it back after dark. This helps balance temperature shifts, creating a stable environment and keeping things cozy without extra energy costs.

You can also think of it like a giant hot water bottle for your plants. Strategically placing thermal mass materials throughout your greenhouse is a smart strategy.

A south-facing **thermal mass wall** is essentially a thick wall made of heat-absorbing materials positioned along the north side of your greenhouse to capture and slowly release heat. (Reverse this if you're in the Southern Hemisphere.)

Also, for a fun research project, look up Trombe Walls. Here's how this clever bit of passive solar engineering works:

- Built from dense materials like concrete, stone, or water-filled containers
- Usually dark-colored to maximize heat absorption
- Typically 8–16 inches thick (thicker walls store more heat)
- Positioned on the north wall to face south through your greenhouse glazing
- Include reflective north wall surfaces for bouncing heat

Think of a thermal mass wall like a rechargeable heat battery: during the day, it soaks up sunlight streaming through your greenhouse, then slowly releases that stored warmth throughout the night, helping maintain steady temperatures. If you wanna test this concept in an affordable way, try placing several black-painted water barrels along your north wall.

Thermal mass also helps during summertime! During hot days, that same heat-absorbing capacity actually helps moderate temperature spikes by soaking up excess heat that would otherwise cook your plants. Here's the catch: for summer cooling, you'll want to help that thermal mass release its stored heat overnight. That means good ventilation when temperatures drop.

If you have water features in your thermal mass strategy, here's a bonus: water not only stores and releases heat effectively, but it also adds some evaporative cooling to the mix. It's like getting a two-for-one deal on temperature control!

Insulation strategies can be utilized to keep your greenhouse warm in winter. Think of this as giving your greenhouse a winter wardrobe—multiple thin layers work better than one thick one. Start with bubble wrap insulation applied directly to the interior walls. Yes, that same bubbly stuff you love to pop can double as thermal curtains for your greenhouse. Simply cut it to size and attach it to your greenhouse walls during winter. If you want the fancier stuff, there is greenhouse-grade insulation available. Less fun and thrifty, but does work better.

Next, add a layer of thermal curtains that can be drawn at night, creating an insulating air gap. For the finale, install a thick insulation panel on your north wall where you get zero winter sun anyway.

On the cooling side, **strategic shade management** is another way of passively managing summer heat. It's like giving your greenhouse a beach umbrella. Use shade cloth strategically positioned on the outside of your greenhouse (key point: outside, not inside!) to block solar gain before it enters. For best results, focus on covering the south and west faces during the hottest parts of the day.

Proper **ventilation** is crucial for temperature control and also for allowing your greenhouse to breathe. It helps with strengthening plant stems, preventing disease, and ensuring proper pollination and transpiration.

Want to harness the sun's power for cooling? (Yes, you read that right!) A **solar chimney** turns scorching sunlight into your greenhouse's personal cooling system through some clever physics. It's a black-painted vertical shaft extending from your roof. As the sun heats this chamber, it creates a powerful upward draft that pulls hot air out while drawing cooler air in through your lower vents. Think of it as your greenhouse's natural elevator, except instead of carrying people, it's whisking away hot air that your plants don't want hanging around.

Once built, this passive system works completely free, turning those "my greenhouse is a toaster oven" days into a much more comfortable growing environment. But wait, there's more! Solar chimneys *also* help with heating.

Your building materials and strategic placement of your greenhouse will also contribute to passive temperature control. A south-facing orientation maximizes sun exposure. Windbreaks from fences, or trees and shrubs, can reduce heat loss from wind exposure. Deciduous trees around a greenhouse can provide seasonal shade during summer and allow sunlight through in winter.

Other passive temperature control techniques include: double glazing or insulated panels, partially buried greenhouses (harness the earth's power), compost piles (which generate heat) inside or nearby, ground heat storage (underground pipes), use of mulches over soil (mulch can moderate temperatures), raised beds or shelving (elevated level improves airflow), and various evaporative cooling methods.

Active Temperature Control

You can absolutely run a greenhouse using only passive methods (plenty of gardeners do!). With careful planning, you can harness things like natural sunlight, airflow, and thermal mass to maintain a stable climate year-round.

However, adding some active temperature control is like upgrading from a bicycle to an e-bike—you'll still need to pedal, but you've got backup power when the hills get steep.

Active systems give you that extra insurance against weather extremes and help maintain more consistent growing conditions, especially if you're pushing seasonal boundaries or growing temperature-sensitive plants. Consider them your greenhouse's safety net, stepping in when passive methods need a helping hand during those polar vortexes or heat waves that make your local meteorologist excited.

Active Warming Methods (For Cooler Seasons)

1. **Electric or gas heaters:**
   - Use portable or mounted heaters to maintain consistent warmth. Models with thermostats and timers provide better control.
2. **Infrared heaters:**
   - Ideal for spot heating, these warm objects (like plants and soil) rather than the air, improving energy efficiency.
3. **Heating mats or cables:**
   - Place under trays or in the soil to keep the root zone at optimal temperatures for germination and growth.
4. **Heat lamps:**
   - Target specific areas with focused warmth for plants needing extra care in cooler weather.
5. **Hot water heating systems:**
   - Circulate heated water through pipes or tubing along greenhouse benches or floors for even warmth distribution.

Active Cooling Methods (For Warmer Seasons)

1. **Exhaust fans:**
   - Draw out hot air while pulling in cooler air from vents or outside, creating a steady airflow.
2. **Evaporative coolers:**
   - Use water-soaked pads and a fan to lower the air temperature through evaporation, perfect for dry climates.
3. **Misting systems:**
   - Release fine water droplets to cool the air and raise humidity, helping plants thrive in peak summer heat.
4. **Air conditioning units:**
   - For ultimate temperature control, use a portable or built-in AC system, though this is energy-intensive.
5. **Shading systems with motors:**
   - Install automated shade cloths or louvers that adjust based on the intensity of sunlight.

If you're up for the investment, automated vent openers are worth every penny. They are temperature-activated for consistent control, and they reduce daily monitoring needs. Plus, they help with humidity regulation. Just my two cents.

If you're *really really* up for the investment, there are environmental controllers available that coordinate multiple systems (fans, heaters, misting, grow lights, irrigation, etc), and can be monitored remotely via smartphone. By automating processes, these controllers reduce energy and water waste, making them environmentally and financially friendly over time.

Pro tip: Start simple with a basic thermostat-controlled fan and heater combo. You can always add more sophisticated systems as you learn your greenhouse's personality and your plants' specific needs. Remember, it's not about having the fanciest setup—it's about maintaining happy plants without becoming a full-time climate manager!

HUMIDITY MANAGEMENT: KEEPING THINGS JUST STEAMY ENOUGH

Think of humidity as your greenhouse's mood—too high or too low, and things get dramatic.

Humidity, that invisible blanket of moisture in the air, plays a crucial role in plant health. It impacts transpiration, the process where plants release water vapor, much like how we sweat. This process helps them stay cool and absorb nutrients. But too much moisture can lead to a jungle of problems, quite literally. High humidity invites mold, mildew, and other unsavory guests who thrive in damp conditions, ready to throw a fungal fiesta on your plants.

Most plants prefer 50–70% humidity (with 60–65% being a general sweet spot), but some like it steamier (tropical plants), and some like it drier (cacti and other desert plants).

Getting your atmosphere right takes some finesse. Let's talk tactics.

When your atmosphere is too dry, you need:

Low Humidity Solutions

- Humidity trays filled with pebbles and water (lower tech)
- Automated misting system with timer (higher tech)

- Damping down paths in the morning on hot days—think "morning dew" (doubles as evaporative cooling)

Or, when things get a little too steamy:

High Humidity Solutions

- Ventilation fans (oscillating fans can work great)
- Dehumidifier for especially swampy days or small spaces
- Strategic morning watering (so leaves dry by night)
- Proper plant spacing for air circulation

Monitoring Tips:

1. Check humidity levels morning and evening.
2. Watch for warning signs:
    - Condensation on walls = too humid
    - Crispy leaf edges = too dry
    - Mold growth = definitely too humid
3. Adjust ventilation before using mechanical solutions.
4. Remember: humidity needs change with seasons.

Pro tip: If you see condensation dripping from the ceiling onto plants, increase air circulation immediately. Nothing invites fungal problems quite like water droplets camping out on leaves overnight.

Remember, all these systems work together—when you adjust one, you'll likely need to tweak the others. Think of it as conducting a tiny orchestra where every instrument needs to play its part just right. Once you get the hang of it, maintaining the perfect environment becomes second nature.

VENTILATION: KEEPING THE AIR MOVING

Think your plants can thrive in stale air? About as well as you'd do in a stuffy elevator. Good ventilation isn't just a nice-to-have—it's essential for healthy plants. Here's why and how to get it right.

Why Ventilation Matters

- Prevents disease by reducing humidity and condensation
- Strengthens stems through gentle air movement
- Helps pollination
- Regulates temperature
- Ensures proper CO2/oxygen exchange for photosynthesis

Types of Ventilation Systems

1. **Natural ventilation**
   - **Roof vents:** Hot air rises and escapes.
   - **Side vents:** Fresh air enters, vents on opposite sides allow for cross-ventilation.
   - Louvered windows
   - *Best for:* Small greenhouses, mild climates
   - *Pro:* No energy costs
   - *Con:* Less control on super hot/cold days
2. **Mechanical ventilation**
   - **Exhaust fans:** These actively remove warm, stale air.
   - **Circulation fans:** These keep air moving, reducing stagnant spots where mold likes to party.
   - Intake shutters, automated vent openers
   - *Best for:* Larger greenhouses, extreme climates
   - *Pro:* Precise control
   - *Con:* Uses electricity, needs maintenance

Pro tip: While your vents do the heavy lifting, solar chimneys and shade cloths act like your greenhouse's backup dancers—they enhance airflow by creating temperature differentials that naturally drive air movement.

Setting Up Proper Airflow

1. **Basic setup:**
   - Place intake vents low on one end.
   - Put exhaust vents high on the opposite end.
   - Install circulation fans for even distribution.
   - Keep paths clear for air movement, and make sure vents are not blocked by any objects.

2. **Fan placement:**
   - Mount circulation fans at plant height.
   - Point slightly downward for best air mixing.
   - Space fans every 20–25 feet.
   - Ensure they can oscillate (stationary fans create dead spots).

Pro tip: Can't decide on fan size? Here's a handy rule—your fans should be able to exchange all the air in your greenhouse every minute in summer, and every 3–4 minutes in winter.

Here's an example to demonstrate what that means. If your greenhouse is 200 square feet and 8 feet tall, you've got 1,600 cubic feet of space. That means you'll need fans that can move 1,600 cubic feet of air per minute (CFM) to handle those summer heat waves.

The math still not clicking? Just multiply your greenhouse's length × width × height to determine your total cubic feet, and that's your target CFM for summer ventilation.

Seasonal Adjustments

- **Summer:**
   - Open all vents early morning.
   - Run fans continuously during peak heat.
   - Consider adding shade cloth for extra cooling.
   - Monitor for signs of heat stress.
- **Winter:**
   - Reduce ventilation but don't eliminate it.
   - Open vents briefly during sunny midday hours.
   - Run fans on low to prevent cold spots.
   - Watch humidity levels (cold air holds less moisture).

Signs You Need Better Ventilation

- Condensation on walls/ceiling
- Still, stagnant air
- Disease problems
- Leggy plant growth
- Temperature spikes on sunny days

Final Thoughts

Good ventilation is like giving your plants their daily exercise—it keeps them strong, healthy, and ready to produce. A little air movement goes a long way in keeping the plant doctor away, and promoting lush, vibrant growth year-round!

It's important to add your ventilation system to your maintenance checklist. Clean fans and vents regularly to prevent clogs and inefficiencies.

And hey, if all this info feels like it's piling up too quickly and too muchly—take a moment, go get some of your own ventilation and give yourself a few deep breaths of fresh air. It's easy to overcomplicate things (ask me how I know). Sometimes, simple vent placement and *maybe* a few well-placed fans are all you need.

Really, learning these different systems just takes some practice. The more you do it, the more you get the hang of it.

LIGHTING SOLUTIONS: BECAUSE YOUR PLANTS DESERVE A SPOTLIGHT

Of course, what would your Broadway production be without lights? In the world of plants, light means life.

So it's the middle of winter, and the sun has decided to take a vacation. We humans might be able to ride out the gloom with hot cocoa and streaming back-to-back episodes. Our green friends, however, haven't quite mastered this level of hibernating. Plants need that golden light to thrive, even when the sky is a perpetual gray.

Enter artificial lighting—your personal sun-in-a-socket! Not only does this supplemental sunshine keep your greenhouse from turning into a plant retirement home, but it's also your ticket to thumbing your nose at seasonal growing restrictions.

Supplemental lighting extends the growing season and gives your plants the energy they need to flourish during shorter days. By mimicking that summer sunshine, these lights keep your plants' energy factory (a.k.a. photosynthesis—the process of converting light to energy) running at full tilt, even when natural daylight is being suspiciously stingy. And those supposedly "winter-shy" bloomers? They may get brave and start showing off, flowering and fruiting like they've forgotten what season it is.

Choosing Your Lights

Let's break down your lighting options:

LED lights

- **Description**: Energy-efficient and long-lasting, LEDs offer customizable spectrums and emit minimal heat, reducing risks to plants.
- **Best for**: General use, energy-conscious growers, and temperature-sensitive plants. Great initial investment that will save you money in the long run.

Fluorescent lights

- **Description**: Affordable and gentle, fluorescent lights provide broad-spectrum light, making them ideal for seedlings and young plants.
- **Best for**: Seed starting, seedlings & young plants, low-light areas, and compact setups.
- **Note: Compact Fluorescent Lights (CFLs)** are smaller versions, easy to install and move, and can be used in smaller greenhouses or as supplemental light for individual plants.

High-Pressure Sodium (HPS) lights

- **Description**: Known for their intense, warm spectrum, HPS lights promote flowering and fruiting but generate significant heat.
- **Best for**: Mature plants in their reproductive stage.

Metal Halide (MH) lights

- **Description**: Emit a cool, blue light that encourages leafy growth and is ideal for vegetative stages.
- **Best for**: Early growth stages of plants.

Incandescent lights

- **Description**: Low-cost but inefficient and heat-intensive, with limited light spectrum for plants.
- **Best for**: Spot treatment in small areas, though rarely recommended for full greenhouse use.

Plasma grow lights

- **Description**: Emit a full spectrum of light that mimics natural sunlight, offering excellent results but at a higher cost.
- **Best for**: Advanced growers with larger budgets.

Setting Up The Lights

Setting up your lighting system is akin to positioning your plants on a stage. Lights need to be positioned for optimal coverage, ensuring every leaf and stem gets its fair share of the limelight.

Installing adjustable fixtures allows you to move lights closer to or farther from plants as they grow (they grow up so fast!), ensuring that light intensity remains just right.

Seedlings are delicate and can scorch under intense light, so you'll start with the lights higher to avoid overwhelming them. As they grow, lower the lights gradually—like easing a kid into swimming lessons instead of tossing them into the deep end. Once they stretch close to the light, you'll raise the fixture just enough to prevent leaf burn while keeping that sweet spot of intensity.

And please, for the love of photosynthesis, get a timer. Because neither you nor your plants want to rely on your memory to maintain consistent light schedules. Timers are your backstage crew in this show, automating light exposure to mimic natural daylight cycles. This consistency helps regulate plant growth, preventing stress from erratic light patterns. Just set it, and your plants bask in the perfect amount of light, day in and day out.

Remember, we're supplementing here. Growing with artificial lights isn't about recreating the sun (talk about pressure!). It's about giving your plants enough light to thrive when natural sunshine is playing hard to get. So you can rest and know that trial and error is okay (and expected). Getting levels right might take some adjusting, but that's perfectly normal.

Detailed Game Plan

- **Height and coverage**
  - Position lights 24–36 inches above mature plants (adjust based on your specific light type).
  - For seedlings, start at 36–48 inches and lower as they grow.
  - Ensure even coverage: one light per 4x4 foot area is a good rule of thumb.
- **Installation tips**
  - Install adjustable chains or pulleys for easy height adjustment.
  - Use light rails if you're feeling fancy (they move lights automatically).
  - Consider investing in reflectors to maximize light distribution.
  - Always secure lights properly—gravity is not your friend here.
- **Timing is everything**
  - Install timers to maintain consistent light schedules.
  - Most plants need 14–16 hours of light during growing season.
  - Allow 8 hours of darkness for proper plant rest (yes, plants need beauty sleep too).

Maintenance and Monitoring

Keep your lighting system running smoothly with these key practices:

- **Daily checks**
  - Watch for any flickering or dim bulbs.
  - Watch for signs of plant stress (leaf burn or stretching).
  - Adjust height as plants grow.
- **Monthly tasks**
  - Clean bulbs and reflectors (dust literally steals the spotlight, and can wear down bulbs faster).
  - Check all connections and tighten if needed.
- **Seasonal adjustments**
  - Increase artificial light during shorter winter days.
  - Reduce during summer when natural light is abundant.
  - Adjust timing to complement natural daylight.

Pro tip: Rotate plants for even growth. Think of young plants as solar-powered dancers—they'll twist and stretch toward light with surprising determination. Give them a quarter turn regularly (yes, like spinning a lazy Susan) to keep growth balanced.

For most setups, rotating plants every one to two weeks is sufficient. Although, young plants can benefit from more frequent rotation since they're still learning how to grow straight and strong.

Your visual clues: upright, balanced growth and evenly distributed foliage with uniform leaf coloration mean you're nailing it. While lopsided growth, leaning plants, or leggy stems are your cues to step up the rotation game.

Signs Your Lighting Is Working (Or Not)

- **Good Signs**
    - Sturdy, compact growth
    - Vibrant leaf color
    - Strong stems
    - Consistent flowering/fruiting
- **Warning Signs**
    - Leggy, stretched growth (needs more light)
    - Leaf burn (too much light or too close)
    - Pale leaves (insufficient light)
    - Slow growth (generally needs more light)

Budget-Friendly Tips

Because who doesn't love saving money while growing money (plants)?

1. Start with fluorescent lights for seedlings.
2. Upgrade to LEDs one section at a time.
3. Use timers to optimize energy usage.
4. Position plants to maximize natural light.
5. Keep your lights clean for maximum efficiency.

Remember: perfect lighting doesn't happen overnight (pun intended). It's okay to start simple and upgrade as you go. Your plants won't judge you for beginning with basic setups—they're pretty grateful just to have a consistent light source. Just avoid the temptation to recreate the Las Vegas Strip in your greenhouse. Sometimes, less is more, especially when it comes to your electricity bill.

Check-In

It's that time again! This chapter was a big one, so let's take stock of ALL that progress you've just made (and there's a bunch):

- You understand the fundamental climate factors that transform a simple structure into a thriving ecosystem.
- You can set up and position monitoring systems to be your greenhouse's early warning system.
- You've learned both passive and active temperature control strategies to keep plants in their "Goldilocks zone".
- You can implement proper ventilation that keeps air moving, strengthens stems, and prevents disease.
- You understand humidity management and how to keep it just steamy enough for your plants.
- You can choose and position supplemental lighting to extend your growing season into the darker months.

… plus a lot more.

These skills are the difference between a greenhouse that's just a glass box and one that's a carefully controlled environment where your plants can thrive year-round.

Let's see where your climate planning stands:

- Have you identified your region's specific climate challenges? (Those quirks that make your local gardening unique)
- Have you planned your primary temperature control strategy? (Passive methods first, active as backup)
- Have you determined how you'll monitor conditions when you're not there? (Since your plants can't text you when they're too hot)
- Have you considered your ventilation approach? (Start with vents, then potentially add fans)
- Have you thought about your lighting needs for darker seasons?

Not quite there on some points? That's perfectly fine! These systems often evolve as you understand your greenhouse's personality more and more. Start with the basics and build from there.

Remember, mastering climate control is a journey, not a race. Take notes on what you're unsure about—we'll tackle those challenges together as we go.

Quick success tip: Create a simple "climate diary" tracking temperature and humidity at different times of day for your first few weeks, and then at infrequent but regular points in coming seasons. These observations will reveal your greenhouse's unique patterns and help you make smarter decisions about when to ventilate, shade, or supplement heat—saving you both energy costs and plant casualties.

Wrap-Up

Just like that, you're officially ready to direct your own climate-controlled production! You've lit the stage, set the atmosphere, and you've got all the basics down for keeping your leafy performers happy. Sure, there might be a few dramatic moments along the way—maybe a diva gardenia having a humidity-related meltdown—but you're learning how to handle it. Take a bow, maestro, you deserve it!

But what about water, you ask? Well, water management is such a crucial part of greenhouse success that it deserves its own chapter. Coming up next, we'll explore everything from basic irrigation to clever water conservation tricks so your plants will never go thirsty again.

5

WATERING WISDOM

TOOLS, TECHNIQUES, AND TIMING FOR A THRIVING GREENHOUSE

You're standing in your greenhouse, watering can in hand, staring at your beloved plants like a nervous parent on the first day of school. "Too much? Too little?" you wonder, as if expecting your tomatoes to pipe up with their drink order. (*"I'll have a light sprinkle, hold the chlorine, thanks!"*)

If this sounds familiar, welcome to the club! Watering might seem simple—just add water, right? But anyone who's ever turned their soil into accidental soup (guilty!) or created an impromptu desert (also guilty!) knows there's more to it than that. It's like making the perfect cup of tea—timing, temperature, and quantity all matter. Get it wrong, and you've got yourself a bitter disappointment. Get it right, and … well, that's why we're here!

UNDERSTANDING INDIVIDUAL PLANT NEEDS

Watering is not a one-size-fits-all activity. It is as diverse as the plants themselves. Different species have distinct preferences, much like people at a potluck.

Take succulents, the camels of our plant party. These water-hoarding champions can thrive on minimal sips, thanks to their clever built-in storage system (those chubby leaves aren't just for show). Meanwhile, ferns are the high-maintenance socialites of the greenhouse, swooning dramatically at the first hint of thirst.

Watering needs are generally available online or on the back of seed packets. You can group plants based on watering needs to make watering trips easier.

The growth stage also plays a role. Seedlings, much like toddlers, need frequent attention, whereas established plants can handle a bit more independence.

Your plants' thirst levels also depend on the type of soil they're planted in. Sandy soils drain quickly, making them the perfect partner for drought-tolerant plants, while clay soils hold onto moisture, ideal for those that prefer their feet wet.

Your greenhouse's environmental conditions significantly modify these baseline watering needs. Higher temperatures accelerate both evaporation and plant transpiration (both processes where water is lost), which means more frequent watering. Elevated humidity levels slow moisture loss through leaves. Therefore—cooler, more humid environments allow for less watering.

As we discussed in the last chapter, keeping your greenhouse climate at an ideal range should actually help your watering needs be more predictable.

The bottom line: Different plants have different drinking habits, so get to know each plant's specific watering needs based on its type, stage of growth, soil, and your greenhouse conditions. It's like being a bartender who remembers everyone's favorite drink, and the tips you get are home-grown produce.

Pretty soon, you'll be reading your plants' signals like ancient weather patterns—catching those subtle leaf droops before they turn into full-blown plant dramatics. Soon, that "is it thirsty or drowning?" guessing game will transform into a confident dance between you, your watering can, and your grateful green friends. And when your neighbor's jaw drops at your thriving cucumber vines? That little smile creeping across your face won't just be pride—it'll be the quiet confidence of someone who speaks fluent Plant.

THE SECRET LANGUAGE OF THIRSTY PLANTS

Let's chat about what your plants are trying to tell you. While it's not quite talking tomatoes, your plants do actually communicate with you. You just have to learn their language.

Reading the Signs

See, your plants are constantly sending you signals about their water and other needs. Here are a few you might encounter:

Too much water:

- Mushy stems or roots (you know how we get pruny fingers when we've been in the bath too long?)
- Mold on the soil surface (nature's way of saying "Whoa there, Captain Splash!")
- Fungus gnats having a pool party

Too little water:

- Crispy, dry leaf edges (their version of chapped lips)
- Slow growth (they're conserving energy)
- Soil pulling away from pot edges (it's not trying to escape; it's just really thirsty)

If your plants could talk, they'd probably say they prefer being a little thirsty to drowning. Kind of like people at a party—better to wish you had one more drink than to regret having too many!

The best approach is to avoid both extremes by understanding your plant's needs, maintaining a consistent watering routine, and finding that sweet spot. But (a little) underwatering is typically easier to fix.

Just like learning any new language, you learn to speak plant with time and practice. You can generally tell when your plant is telling you something, or when something looks or feels "off."

Sometimes, one sign can point to multiple issues. Yellowing leaves, for instance, can mean multiple things, including both underwatering and overwatering. Also: nutrient deficiency, temperature stress, light issues, pests, diseases, soil pH, or plain ol' natural aging. Suffice it to say, unless your plant is naturally yellow (like sunshine ligustrum), you should slow down when you see that "yellow light" in your greenhouse.

So how do you know what the yellow is for? As you learn the language, you learn to check for other clues to help you narrow down. For example, yellowing due to overwatering is often accompanied by soggy soil and limp, mushy leaves where the yellow is widespread and uniform. Yellowing due to underwatering comes with light, dry soil, and papery, crisp leaves where yellowing often starts at the edges and curls.

Here's another example of listening closely to your plants. Say your plant is wilting, but the top of the soil is still moist. This could mean incorrect or uneven watering. It's important to water deeply and evenly, until water runs out of drainage holes. So here's how you dig deeper (literally) for more clues: check soil moisture at different depths.

How do you do that? Like so:

The Finger Test: Your Built-in Moisture Meter

Before you invest in fancy equipment, remember: you've got five perfectly good moisture meters attached to your hand. The finger test is like the original smart technology—just stick your finger about an inch into the soil.

- **Just Right:** Feels like a wrung-out sponge? Perfect! Moist soil feels cool, slightly damp, and may leave some residue on your finger, but shouldn't be muddy. *Your plants are adequately hydrated.*
- **Too Wet:** Feels like cake mix? Too wet! Waterlogged soil feels soggy or muddy and may cause water to seep into the hole you created. The soil might feel paste-like and stick heavily to your finger. *Hold off on watering and investigate drainage.*
- **Too Dry:** Feels like your forgotten granola? Too dry! Soil feels powdery or crumbly and may not hold together when pressed. In severely dry conditions, it might feel hard or crusty on top. *Time to water thoroughly.*

An Actual Soil Moisture Meter

While the finger test is reliably handy, a soil moisture meter brings some serious perks to your greenhouse game. For starters, it can probe much deeper than your finger, reaching down to where the real root action happens. This becomes especially valuable when you're growing deep-rooted plants like tomatoes or managing multiple containers with different watering needs.

Plus, for those finicky plants that demand precise moisture levels (I'm looking at you, orchids), a meter really levels up your watering game. Instead of guessing and hoping for the best, you get straightforward numbers that tell you exactly when it's time to water—and when it's definitely not.

Think of it as upgrading from a rough estimate to GPS-level precision in your watering routine—particularly useful when you're juggling dozens of plants with different moisture preferences under one roof. Great to have in your back pocket (actually) for quick, precise reads.

WATERING LIKE A PRO: TIMING, TECHNIQUE & TIPS

With just a few simple tweaks, you can level up your greenhouse watering routine to pro-level plant-nourishing—small changes, big results!

Timing Is Everything

So when's the best time to water?

Answer: Early morning.

I mean, who doesn't like breakfast in bed? Plants have all day to drink up that moisture and dry their leaves before bedtime. Dry leaves by evening means less chance of fungal growth. Early morning watering also reduces evaporation, so your plants can soak up every drop.

Next best? Late afternoon, if mornings are hectic.

Evening watering should be avoided if possible. It's like a midnight snack; occasionally a necessity, but not good for the waistline (or, in this instance, the rootline). Evening watering can lead to mildew and fungal growth.

Avoid watering at noon, too. This is mainly because midday watering can lead to increased evaporation and is generally less efficient, with less water actually reaching those thirsty roots.

Some sources also say that watering at high noon is like giving your plants a magnifying glass treatment. Those water droplets on leaves can actually focus sunlight like tiny lenses, potentially scorching your plants. Ouch!

Well, even if it's just to make your watering go further, it's best to save the hot lunch date for yourself, not your plants.

Some Golden Rules of Watering

1. **Water deeply but less frequently**
   - Think of it as training your plants' roots to be treasure hunters rather than surface skimmers. Deep watering encourages them to dig down and explore for their gold (water), building a stronger foundation in the process.
2. **Watch your water temperature**
   - Would you enjoy an ice-cold shower first thing in the morning? Neither do your plants!
3. **Direct water to the roots**
   - Leaves are for photosynthesis, not for swimming.
4. **Check moisture before you pour**
   - "But I always water on Wednesdays!" isn't a valid excuse if the soil is still wet.
5. **Adjust watering schedule with the seasons**
   - While a greenhouse provides a controlled environment, seasonal changes still influence factors like temperature, humidity, and sunlight, all of which affect how much water your plants need.
   - More frequent watering may be necessary during spring and summer, while less is needed in autumn, and even less in winter.

FANCY UPGRADES TO YOUR WATERING TOOLKIT

In Chapter 3, you learned about the essential watering tools in our handy Master Tool Guide.

Here's a review of the basics: a watering can, garden hose with adjustable nozzle, soil moisture meter, and spray bottle.

But now, maybe you've got the watering basics down, but you've got the itch to take your watering game from manual transmission to automatic pilot. Systems like automatic irrigation, misting systems, and rainwater collection are an investment, but definitely worth it.

These systems may cost more upfront, but think of it as investing in your garden's future (and your own peace of mind). Automated irrigation can slash your water usage by up to 30% through precision timing and targeted delivery. Plus, consistent watering leads to healthier plants with stronger root systems and better resistance to diseases and pests.

You'll also free up time for other garden tasks—or maybe just enjoying your morning coffee while admiring your thriving plants instead of lugging watering cans around. And let's not forget the vacation factor: no more desperately trying to teach your neighbor the difference between "slightly moist" and "soaking wet" while you're away. Your plants get exactly what they need, when they need it, whether you're there or not.

AUTOMATED IRRIGATION: SET IT, FORGET IT, LET IT FLOW

Let's talk about one of the smartest investments you can make in your greenhouse: automated irrigation. Think of it as hiring a highly efficient assistant who never forgets to water your plants, takes weekends off, or gets distracted by cat videos on the internet.

Yeah, there are some setup hurdles to jump. And you might just find yourself accidentally creating a sprinkler system that's pointed at your walls (ask me how I know). But after some frustrated grunts, some laughs, and tweaking your system til you get it right, you'll be really glad you did this and persevered.

Why Automate Your Watering?

Automated irrigation isn't just about convenience (though believe me, that's a fantastic perk). It's about precision, efficiency, and giving your plants exactly what they need, when they need it. Here's what makes it worth considering:

- **Consistency**: Plants thrive on routine, and automated systems deliver water with clockwork precision.
- **Water Conservation**: Targeted delivery means less waste through evaporation or overwatering.
- **Time Management**: Free yourself from daily watering duties without compromising plant care.
- **Plant Health**: Direct root-zone watering reduces leaf diseases and promotes stronger growth.

Types of Systems

In the world of irrigation, different systems cater to different needs.

Drip irrigation shines in targeted watering, perfect for greenhouses where individual plant attention is key.

If you're working with a larger area, sprinkler systems provide broader coverage, though less precision.

It's like choosing between a detailed paintbrush and a roller—both have their place, depending on the canvas.

And sometimes, using both is a great idea.

Let's take a look at a few types:

1. **Drip irrigation systems**
   - **How it works**: Water is delivered in a slow, steady stream, straight to the base of your plants (root zone) using a network of tubes and small emitters.
   - **Advantages**:
     - Highly efficient water usage.
     - Reduces water runoff and evaporation.
     - Ideal for plants with specific water needs, and container plants.
   - **Considerations**:
     - Requires regular maintenance to prevent clogging.
     - Initial setup can be intricate.
2. **Overhead sprinkler systems**
   - **How it works**: Water is distributed through overhead sprinklers, mimicking natural rainfall. It's like traditional lawn sprinklers for your greenhouse.
   - **Advantages**:
     - Covers large areas efficiently, for large, uniform growing areas.
     - Suitable for seedlings or plants that require foliar moisture.
   - **Considerations**:
     - Can lead to water wastage through evaporation.
     - May promote fungal diseases if foliage remains wet for long periods.

3. **Misting or fogging systems**
   - **How it works**: Often used to regulate humidity. Releases fine water mist or fog to increase humidity and water-sensitive plants.
   - **Advantages**:
     - Excellent for propagating cuttings, tropical plants, and maintaining high humidity.
     - Reduces water usage compared to sprinklers.
   - **Considerations**:
     - Requires precise control to avoid over-humidifying the greenhouse.
     - May not provide adequate root watering.
4. **Smart irrigation systems**
   - **How it works**: Uses sensors and automation to monitor soil moisture, weather conditions, and plant needs, and adjusts watering schedules dynamically.
     - *Note: Any of the above systems can be made into smart systems.*
   - **Advantages**:
     - Maximizes efficiency with real-time adjustments.
     - Reduces water waste and overwatering.
   - **Considerations**:
     - Higher upfront costs.
     - Dependent on reliable power and technology.

These systems can be used on their own, or together. Each system has its niche, so they can be combined for your desired effect. Let's zoom in on one in particular:

DRIP IRRIGATION: THE LAZY GARDENER'S BEST FRIEND

Drip irrigation is the MVP of automated watering systems. It delivers water directly to plant roots through a network of tubes and emitters. Think of it as a carefully planned subway system for water, with stops exactly where your plants need them.

Setting up a basic drip system requires a moderate level of skill, but you can absolutely do it! I'll give you a basic overview.

WATERING WISDOM

Components You'll Need:

- Main water line (the central pipeline)
- Distribution tubing (smaller lines that branch off)
- Drip emitters (the points where the water is delivered)
- Filter (keeps debris from clogging the system)
- Pressure regulator (maintains consistent flow)
- Timer (automates the whole operation)
- Backflow preventer (protects your water supply)

Setting Up Your System

1. **Planning Phase**
   - Map out your greenhouse layout.
   - Calculate water needs for different plant zones.
   - Determine emitter placement and flow rates.
   - Plan for future expansion.
2. **Installation Steps**
   - Install the main water line along your greenhouse spine.
   - Connect distribution tubes to reach plant zones.
   - Place emitters near plant root zones.
   - Install filter and pressure regulator at the water source.
   - Set up your timer and test the system.
3. **Customization Tips**
   - Use different flow-rate emitters for varying plant needs.
   - Install shut-off valves to control different zones.
   - Consider adding a fertigation system for automated fertilizing.
   - Include moisture sensors for smarter water management.

Fine-Tuning Your System

The key to success is matching water delivery to your plants' needs:

- **Morning Glory Plants** (heavy drinkers):
  - Multiple emitters per plant
  - Higher flow rates
  - More frequent watering cycles

- **Succulent Squad** (light sippers):
  - Single emitter per plant
  - Lower flow rates
  - Less frequent watering cycles

Maintenance Checklist

Regular maintenance keeps your system running smoothly:

- **Weekly tasks:**
  - Check for clogged emitters.
  - Monitor plant response.
  - Adjust timing if needed.
- **Monthly tasks:**
  - Flush the system.
  - Check for leaks.
  - Clean the filter.
- **Seasonal tasks:**
  - Deep clean the system.
  - Adjust watering schedules.
  - Check for winter protection needs.

Troubleshooting Common Issues

1. **Uneven water distribution:**
   - Check for clogs in emitters.
   - Verify pressure regulation.
   - Look for kinked tubing.
2. **System not running:**
   - Check power to timer.
   - Verify water source is on.
   - Inspect for major leaks.
3. **Plant stress despite irrigation:**
   - Test emitter flow rates.
   - Check soil moisture levels.
   - Adjust watering duration.

Pro tips for Success

1. **Zone planning**
    - Group plants with similar water needs.
    - Account for sun exposure differences.
    - Plan for seasonal changes.
2. **Water conservation**
    - Water early morning or late afternoon.
    - Use mulch to reduce evaporation.
3. **System protection**
    - Use UV-resistant tubing.
    - Install flush valves at line ends.
    - Consider adding a water meter to track usage.

Remember: The goal isn't simply to automate watering—it's to create a system that delivers the right amount of water to each plant while conserving resources and simplifying your gardening routine. With proper setup and maintenance, your automated irrigation system will feel less like a luxury and more like an essential greenhouse team member.

Budget-Friendly Tips

Starting small is perfectly fine. Here are some basic tips:

1. Begin with basic timer and main line.
2. Add automation features gradually.
3. Expand zones as needed.
4. Use manual valves before investing in electronic ones.

Your greenhouse will thank you with happy, well-hydrated plants, and you'll thank yourself every time you go on vacation without worrying about crispy leaves greeting you upon return.

RAINWATER HARVESTING: BECAUSE FREE WATER IS THE BEST WATER

And while we're on the subject of water, why pay for it when the sky gives it away for free? Setting up a rainwater collection system is like putting out buckets during a sale at the water store—except this sale happens every time it rains!

Think of it as nature's subscription box service, delivering fresh, chemical-free water right to your greenhouse door. And unlike those other subscription boxes, you'll actually use everything that comes in this one!

Basic Setup

- **Gutters** (the water collectors): Think of these as your greenhouse's version of outstretched arms, ready to catch every drop.
- **Downspouts** (the water slides): Your greenhouse's very own water park features.
- **Storage tanks** (the water bank): Where you stash your liquid savings.
- **Filtration** (because leaves aren't a garnish): Like a bouncer at an exclusive water club, keeping out the riffraff.

Making It Work

Remember playing with those marble runs as a kid? Setting up your rainwater system is similar, except instead of marbles, you're channeling nature's sprinkles toward your thirsty plants. The trick is making it a seamless flow from roof to tank—no water parkour allowed!

Start with good guttering (slightly sloped, like a subtle hint toward the exit at a party), leading to downspouts that can handle your roof's worth of rain during a decent storm. Think of it as designing a water slide: you want the fun (water) to keep moving, not create unexpected pools in your gutters!

For storage, you've got options ranging from modest rain barrels to massive tanks that could double as swimming pools for gnomes.

Choose based on your:

- Rainfall patterns (How generous is your local sky?)
- Garden size (How thirsty are your plant friends?)
- Available space (Where will you park your water savings account?)

Pro tip: Dark-colored or opaque tanks are like sunglasses for your water—they prevent algae from throwing a green party in your storage system.

Keeping It Clean

Remember that time you tried drinking from a garden hose and got a surprise leaf smoothie? Let's avoid that for your plants.

Install:

- Leaf screens (the bouncers)
- First-flush diverters (like skimming the foam off your latte)
- Intake filters (the final checkpoint)

This gives your plants clean drinking water, *and* it prevents clogged pipes.

Making It Legal

Before you go full rain baron, check your local regulations. Some places have rules about rainwater harvesting that are quite complex. Most areas encourage it, but it's better to know the rules than get that awkward "we need to talk" letter from your local authorities.

Bonus Points

Want to level up your rain game? Consider:

- Automated first-flush systems (because who wants to stand in the rain pulling levers?)
- Pressure pumps (giving your water some oomph when gravity needs a helping hand, these ensure a steady flow of water from your storage tanks to your plants.)
- Water level indicators (no more playing "guess how full the tank is.")
- Multiple tank setups (like having backup servers for your water supply.)

Final Thoughts

Harvesting rainwater got you excited, but your wallet got you un-excited? Don't let a tight budget rain on your water-collecting parade! While fancy systems are nice, you can start harvesting rain with something as simple as a clean food-grade barrel under your greenhouse's drip line, paired with some repurposed guttering and a DIY filter made from basic window screening. As usual, there are creative ideas out there, friend.

Keep in mind: Every drop you harvest is one you don't have to pay for later. It's like finding money in your pocket, except it falls from the sky! Plus, your plants will thank you—they're secret rain connoisseurs who prefer their water *au naturel*.

Now go forth and harvest that sky juice! Your wallet (and your plants) will thank you.

TROUBLESHOOTING: WHEN THINGS GO WRONG (AND THEY WILL)

We all make mistakes. I have … you will … Think of them as lessons, and grow from the experience.

Common Oops Moments and How to Fix Them

- **The Swamp Situation (Overwatering):**
  1. Stop watering (obviously) … dial back frequency and quantity.
  2. Improve drainage (add perlite to the soil mix, ensure proper drainage holes in containers).
  3. Consider repotting if things are really soggy (like moving to higher ground during a flood).
- **The Desert Dilemma (Underwatering):**
  1. Gradually reintroduce water (no need to recreate the great flood). Water slowly and thoroughly to avoid shocking the plant. Allow water to seep into the soil evenly.
     - If your soil's drier than a summer sidewalk, try watering from the bottom up—just set the pot in a shallow dish of water and let it drink at its own pace.
  2. Mulch (like putting a blanket on your soil) … Add a layer of organic mulch like straw or compost to help the soil retain moisture longer.
  3. Adjust your watering schedule (set those reminders).

An Ounce of Prevention is Worth a Pound of Cure

Mulching is a simple yet powerful strategy to maintain consistent soil moisture. Adding a layer of organic material like straw or bark insulates the soil against evaporation and temperature fluctuations. This method acts as a cozy blanket for your plants, keeping them snug and hydrated.

Other preventative measures include:

- Learn your plant's specific watering needs.
- Group plants by water needs.
- Check on soil moisture regularly.
- Keep an eye on environmental conditions.
- Conduct routine inspections of your irrigation systems, much like you would check your car's oil levels to prevent a breakdown.

CHECK-IN

Well, would you look at that? You've upgraded from watering randomly to informed hydration. I'll raise my watering can to that! Let's celebrate what you've accomplished so far:

- You can now interpret your plants' thirst signals—from desperate leaf droop to contented moisture levels.
- You understand how different plants have different watering needs—from desert-loving succulents to thirsty ferns.
- You've mastered the famous "finger test"—your built-in soil moisture meter that goes everywhere you do.
- You know the golden rules of watering—including timing, depth, and technique.
- You can evaluate different irrigation options—from simple watering cans to full automation systems.
- You understand rainwater harvesting basics—turning free sky-juice into plant nutrition!

These skills transform watering from a guessing game into a precise science that will keep your plants not just watered, but watered *well*.

Let's see where your moisture management stands:

- Can you identify signs of overwatering versus underwatering in your plants?
- Have you grouped some plants by similar watering needs for efficient maintenance?
- Have you thought up a consistent watering routine that accounts for morning timing?
- Have you chosen your primary watering method and tools?
- Do you have a plan for watering during vacations or busy periods?

Not quite confident about some of these? No problem! That's exactly why we're here. Make note of what you're unsure about, and we'll keep building knowledge and skills together.

Start with mastering the finger test and learning to read your plants' signals—the rest will flow naturally from there.

In fact, that's a great thing to go try right now, or as soon as you can. Try the finger test on three different plants. Write down what you feel and what it tells you about each plant's water needs. This hands-on practice is how you build that watering intuition we talked about!

Quick success tip: You can label each plant with its watering preference using a simple color system — blue tags for moisture-lovers, yellow for moderate needs, and red for drought-tolerant plants. This visual shorthand will speed up your watering rounds and prevent both soggy soil and crispy leaves.

Wrap-Up

Here's the final drip. Remember this: perfect watering is like perfect parenting—it doesn't exist. But good enough watering? That's totally achievable. Your plants are more resilient than you think, and with these guidelines, you're well on your way to finding that sweet spot between desert and swamp.

Now that you've got your watering wisdom dialed in, it's time to tackle the nuts and bolts (quite literally) of putting your greenhouse together. Let's roll up our sleeves and dive into the structural setup of your greenhouse, from choosing the right foundation to creating efficient pathways that'll make your daily plant-tending tasks flow as smoothly as a well-designed irrigation system.

6

HOW TO BUILD YOUR GREENHOUSE

FOUNDATIONS, FRAMING, AND FINISHING TOUCHES

Alright, greenhouse dreamers, it's time for the moment of truth: actually building this thing! Remember when we talked about choosing your greenhouse type and materials? Well, now we're going to turn those plans into reality.

I'll be like your virtual construction buddy—minus having to share your lunch or listen to my terrible dad jokes in person. Together, we'll turn that pile of parts into something amazing.

Don't worry if you're feeling a mix of excitement and mild panic—that's completely normal. This greenhouse dream is your baby! Think of this chapter as your "greenhouse birth plan"—except instead of breathing exercises, we'll be focusing on proper tool use and level surfaces. (Though some deep breathing might come in handy too!)

CHOOSING YOUR FOUNDATION TYPE

They say having a solid foundation really helps keep you grounded. See, there I go with the shameless puns ... Can't help myself! In all seriousness, though, laying a good foundation is a very important part of the process. Without a good one, slowly but surely, it'll all come crashing down like the last round of Jenga.

Not only that, but a good foundation also ensures longevity and shields your precious plants from unwanted guests like burrowing critters and sneaky drafts.

You've got several options here, each with its own personality to fit your goals, climate, and budget. Here are some common options:

1. Bare Earth: The Minimalist Choice

- **Best for**: Small, temporary, or budget-friendly setups. Starting small or keeping your options open.
- **Pros**: Cost-effective and promotes natural drainage (drains better with certain soil types: see Chapter 7). Easy to modify if you change your mind later.
- **Cons**: Susceptible to weeds and less stable for large greenhouses. Can get muddy.
- **Usage tip**: Cover with landscape fabric or gravel to reduce weeds.

2. Gravel Base: The Happy Medium

- **Best for**: Mid-sized greenhouses needing stability and drainage.
- **Pros**: Affordable, excellent drainage, and good for anchoring structures.
- **Cons**: Can shift over time and isn't ideal for heavy greenhouses. Can play hide-and-seek with small tools.
- **Usage tip**: Layer it like a cake: weed barrier on bottom, gravel on top. Your future self will thank you.

3. Concrete Slab: The Forever Home

- **Best for**: Permanent or larger, more heavy-duty greenhouses.
- **Pros**: Extremely durable, permanent, easy to clean, and keeps pests out, and ideal for maintaining consistent temperature (great thermal mass for temperature control).
- **Cons**: Expensive and poor drainage unless designed with drainage channels. Requires permits, might need professional help.
- **Usage tip**: Include drainage systems for excess water. Think of them as water slides for excess moisture.

4. Wooden Frame: The Natural Charmer

- **Best for**: DIY enthusiasts or greenhouses in moderate climates.
- **Pros**: Affordable and easy to assemble, with a naturally beautiful aesthetic.
- **Cons**: Can rot unless treated or raised off the ground, and it needs regular maintenance. Not ideal for wet climates unless properly prepared.
- **Usage tip**: Use treated lumber or rot-resistant wood like cedar.

5. Pavers or Bricks: The Aesthetic Achiever

- **Best for**: Medium-to-large greenhouses with a focus on aesthetics.
- **Pros**: Beautiful, durable, and decent drainage between the cracks.
- **Cons**: Labor-intensive to install (precise and long) and costly for large areas.
- **Usage tip**: Use compacted sand and gravel underneath for stability. It's like giving your pavers a proper mattress to rest on.

6. Hybrid Foundation: The Best of All Worlds

- **Best for**: Greenhouses requiring customized solutions, and the "why not both?" crowd.
- **Pros**: Combine materials (e.g., concrete perimeter with gravel floor) for tailored benefits, adaptable to specific needs, and can save money in strategic areas.
- **Cons**: Complexity can increase costs, and requires careful (more complex) planning.
- **Usage tip**: Use the concrete perimeter to anchor the greenhouse while leaving a more natural floor inside for planting. Consider gravel or soil on the inside, for instance. It's like having a secure frame with a cozy interior—the best of both worlds!

Each foundation type has its ideal applications and trade-offs. Consider factors like soil drainage, long-term durability, the greenhouse's intended use, and even your desired aesthetic to choose the perfect fit for your gardening dreams!

THE LEVEL-HEADED APPROACH TO SITE PREP

Before you lay that foundation, you need to get your site level. Here's your very basic step-by-step:

1. **Clear the area**
    - Remove all vegetation (yes, even that stubborn dandelion).
    - Clear rocks, roots, and debris.
    - Think of it as giving your greenhouse a clean slate to start fresh.
2. **Check for level**
    - Use a long level or string level.
    - Mark high spots and low areas.
    - Remember: water always flows downhill, so a slight slope for drainage isn't bad. A gentle slope of about 1–2% (that's about a 1 to 2-inch drop over 8 feet) can actually be your drainage friend. Think of it like tilting your plate ever so slightly to keep the gravy from pooling—just enough to guide the water where you want it, but not enough to send your peas rolling onto the floor.
3. **Adjust the ground**
    - Fill low spots with appropriate material (gravel, soil).
    - Run a landscaping rake (or something with a flat end like that) back and forth over the surface, and level out the dirt.
    - Tamp down all filled areas firmly.
    - Think of it like making your bed—except this one needs to be really, really flat.

Pro tip: If you're on a slope, you might need to cut into the hill or build up the low side (or try out a nifty invention like that uneven span greenhouse). Just remember: water is that friend who always finds a way to crash the party—plan for drainage!

THE GREAT ASSEMBLY: PUTTING IT ALL TOGETHER

Now comes the fun part—actual assembly! This is when your greenhouse starts looking less like a pile of parts and more like that plant paradise you've seen in your daydreams. (Possibly night dreams too, if you obsess over projects like I do).

Quick side note. I remember many moments staring blankly at project components scattered across my yard like a construction yard sale. Then help showed up. Those moments make me smile.

Sometimes all it takes to ease the overwhelm is a good buddy (or spouse) who can come on the scene, bust out laughing with you (at you?), and roll up their sleeves. Six hours, three scraped knuckles, and a dozen or so "where does THIS go?" later, you might just find yourselves high-fiving beneath the frame.

The moral? A building buddy really does help—even if they spend half the time teasing you about your measuring skills.

The Assembly Game Plan

1. **Lay Out All Parts**
   - Organize pieces by type.
   - Count everything twice.
   - Think of it like preparing ingredients for a recipe—you don't want to realize you're missing something halfway through.
2. **Follow the Frame Up**
   - Start with a perimeter frame, and the base frame assembly.
   - Work your way up with vertical supports.
   - Add cross-bracing where needed.
   - Ensure each connection is tight, and check for alignment and squareness frequently.
3. **Panel Installation**
   - Work systematically from bottom to top.
   - Clean all panels before installing.
   - Ensure proper overlap for water runoff.
   - Think: shingles on a roof, but transparent.

Tips for Building Success

- **Read instructions:** Yes, really. Most kits come with detailed instructions. Even if you're usually a "figure it out as you go" person. Trust me on this one.
- **Check weather:** Choose a calm, dry day. Wind and rain are not your friends during construction.
- **Gather help:** Some parts really do need an extra pair of hands.

- **Double-check measurements:** Measure twice, cut once, and maybe measure a third time just to be sure. Patience is one of your best tools here.
- **Check upcycled material:** If you're using old windows and doors, ensure they are tested for lead paint and adjusted to fit snugly within the frame.
- **Reinforcements:** Anchor your greenhouse to the ground using stakes or bolts to prevent it from doing a Mary Poppins on a windy day. In regions prone to snow or heavy rain, consider additional reinforcement options like cross-bracing or adding extra support beams.

And also:

- Start early in the day.
- Have all tools ready before beginning.
- Keep instructions in a clean, dry spot.
- Take photos as you go (great for remembering how things fit together).

DOORS AND WINDOWS: GETTING YOUR GREENHOUSE TO OPEN UP

Doors and windows aren't just access points—they're strategic openings that control your indoor climate, providing airflow for your greenhouse. Like the gills of a fish, they allow your greenhouse to breathe.

Door Installation Tips

This is your grand entrance for you and your plants, so choose and prepare it well!

1. **Location matters**
   - Position for easy access from your main garden area or home.
   - Consider prevailing winds.
   - Think about your workflow (you'll be going in and out A LOT).
2. **Proper hanging**
   - Ensure frame is square.
   - Use heavy-duty hinges.

- Weatherproof the door with sealant around the edges and weatherstripping.
- Make sure the door is level and swings freely without scraping the ground.
- Pro tip: A slightly sloping threshold helps prevent water pooling.

Window Wisdom

Windows are all about maximizing light and airflow.

1. **Placement strategy**
   - Position for cross-ventilation. Place windows on opposite walls to keep the air moving and prevent hot spots that could stress your plants.
   - Strategically position them to harness prevailing winds and create a natural cross-breeze, allowing fresh air to circulate throughout the space.
   - Consider morning vs. afternoon sun.
   - Think about reaching height for operation.
2. **Types of windows**
   - Manual vs. automatic openers
   - Sliding vs. hinged
3. **Remember ...**
   - Seal the deal with a silicone sealant, which can flex with temperature changes, to make your windows airtight and efficient.
   - Easy access = more likely to use them.

Maintenance

A squeaky door hinge or a sticky window latch might not seem like a big deal, but over time, it can turn into a frustrating obstacle. Regularly lubricate moving parts—WD-40 is your best friend here—and keep an eye out for cracks or warping. If you spot damage, tackle it sooner rather than later to maintain a happy, functional greenhouse.

Remember, doors and windows may not have the glamour of a tomato vine heavy with fruit, but they're vital to your greenhouse's success. So, go ahead and give them the proper installation and attention they deserve.

SAFETY FIRST (BECAUSE NOBODY WANTS A GREENHOUSE MISHAP)

Alright, I know safety isn't the sexiest topic in greenhouse gardening. It's like being reminded to floss—important but not exactly thrilling. It's crucial to not let your enthusiasm outpace your caution here, though, because nothing ruins a garden day quite like an unplanned trip to urgent care.

So, let's talk about making your greenhouse more of a sanctuary than a hazard zone.

Essential Safety Gear *(Hope for the Best, Plan for the Whoops)*

- **Gloves:** Your hands will thank you.
- **Safety glasses:** Because eyes and flying debris don't mix.
- **Closed-toe shoes:** Dropping a panel on flip-flop-clad feet? No thanks.
- **Hard hat:** Dropping a panel on a bare head? Also no thanks.
- **First aid kit:** Keep one handy, hope you never need it.
- **Eye wash:** Because soil has impressive aim sometimes.
- **Emergency contact numbers:** Because panicking and yelling "Help!" isn't exactly a reliable communication plan.

During Construction

- Work with a buddy when possible
- Keep your work area clean
- Take breaks (dehydrated building = wobbly building)
- Listen to your body (and the instructions)
- Always ensure ladders are on level ground before climbing
- Remember, tools are like your teammates—treat them with respect, and they'll perform like champions

Think Future Safety As You Build

- Plan for proper ventilation, to ensure good airflow inside and prevent heat build-up.
- Remember how important structural integrity is when building.

Structural Stability *(Because Flying Greenhouses Belong in Oz)*

- Secure all panels and frames properly—think of it as giving your greenhouse a really good seatbelt.
- Anchor that beauty down like it owes you money. Those ground anchors aren't suggestions; they're your greenhouse's roots.

Electrical Safety *(Because Shocking Developments Should Stay in Soap Operas)*

- Keep all electrical connections high and dry—water and electricity mix about as well as a porcupine and a balloon.
- Use outdoor-rated equipment and GFCI outlets.
- Route cords and cables where they won't turn your greenhouse into an obstacle course.

Chemical Safety *(Because Your Greenhouse Isn't Breaking Bad)*

- Store fertilizers and pest controls in sealed, labeled containers. (We will discuss these more in-depth in later chapters).
- Keep them high and dry, away from curious critters (including the human variety).
- Have a designated spot for safety gear—gloves, masks, and goggles should be easy to find.

The Path Less Hazardous

- Keep walkways clear (your plants don't need to see your interpretive dance moves as you trip over tools).
- Clean up spills fast.
- Good lighting isn't just for plants—you need to see where you're stepping too!

Safety and Maintenance After Building

Regularly check your framework for any signs of wear and tear, much like you would a favorite pair of shoes before a long hike. You want to catch small issues before they become big problems.

Check for loose screws, cracked panels, and any sneaky damage that could spell trouble. Monthly is good, and after storms is also important.

Tools and chemicals should have their designated home, stored safely out of reach of curious hands or paws. Everything in its place means no surprise hazards later.

Be prepared for emergencies: with a well-stocked first aid kit, a list of emergency contact numbers handy, and an evacuation plan for severe weather scenarios.

Pro tip: Establish an emergency shut-off procedure. Put it in writing (weatherproof that sheet of paper!) and leave it where it's easy to see. Trust me, in a panic, even remembering how to turn off a water valve can feel like advanced calculus.

Final Thoughts

We want your greenhouse to be your happy place, not your "I should have known better" place. Taking these precautions isn't being paranoid—it's being smart, like wearing oven mitts to handle hot pans or checking for toilet paper before sitting down.

A little preparation goes a long way in keeping your greenhouse adventures strictly about the joys of growing, not the joys of explaining to your insurance company why your greenhouse is now in the neighbor's yard.

Now that we've covered the serious stuff (while hopefully keeping you awake), you're ready to garden with both enthusiasm AND wisdom.

Because the best kind of gardening is the kind where everyone—including you—stays safe and happy to grow another day!

Check-In

Well, you did it. You actually built it.

Or at least, you know know how to, and you're making solid plans! Remember when that was just a far off dream? And now you're here.

By working through this foundation-to-finish journey, you've accomplished quite a lot. You now:

- Understand the pros and cons of different foundation types, from the minimalist bare earth approach to the "forever home" concrete slab.
- Know the basics on how to properly prepare and level your site.
- Can confidently navigate the assembly process, from organizing parts to following that all-important frame-up sequence.
- Can install doors and windows strategically for proper ventilation and access.
- Know how to implement crucial safety measures to keep your gardening adventure from becoming an urgent care visit.
- Will properly anchor your greenhouse to prevent it from "doing a Mary Poppins on a windy day".

Let's check where your greenhouse construction journey stands:

- Have you selected the foundation type that best matches your climate, budget, and long-term greenhouse goals?
- Did you map out your site with proper consideration for drainage (remember that 1–2% slope—just enough tilt to guide water without sending your peas rolling)?
- Do you have a plan for doors and windows with cross-ventilation in mind?
- Have you got all your materials organized and counted?
- Do you have your safety gear ready?
- Have you read through ALL the instructions? (Be honest!)
- Have you recruited your building buddy?

Again, no problem if you're not quite there on all the points. Better to pause now than realize you're missing something crucial before your start, or even halfway through. Use this as your pre-flight checklist before takeoff.

Quick success tip: Take photos throughout your building process! Not only will they help you troubleshoot if something goes awry, but they'll become treasured before-and-after evidence of your greenhouse journey. Nothing beats the satisfaction of seeing that pile of parts transform into the real deal.

Wrap-Up

Building your greenhouse is like putting together a giant puzzle that will eventually give you tomatoes. Take your time, follow the steps, and remember—if something doesn't feel right, stop and double-check. It's better to spend an extra few minutes making sure everything's correct than to realize later that your greenhouse is doing an impression of the Leaning Tower of Pisa.

Remember: Every expert greenhouse gardener started somewhere, probably with a pile of parts and an instruction manual, just like you. The key is to keep going, take it step by step, and celebrate the small victories—like when you realize all your screws actually went where they were supposed to! Yaaay!

Now that we've talked a good deal about proper building and safety, I think it's high time we bring in this reminder: make sure to have fun with this process! I mean, you're really doing it!

Like I said, take lots of photos for memories, and celebrate milestones (frame up = time for a snack break). Even take snapshots of those roadblock moments, because someday it'll be great to look back on them. Today's headaches will be tomorrow's laughs and success stories.

As you build this greenhouse, you're building your dreams, and creating your own little slice of paradise. It may be hard work, but trust me, that first seedling you grow in there will make all this effort worth it!

Speaking of seedlings, they're going to need somewhere cozy to put down roots. Next up, we'll dig into the dirt on soil and fertilizer—because even plant babies need a proper nursery! Get ready to become a soil sommelier and learn the difference between good dirt and great dirt.

7

SOIL AND FERTILIZATION

LAYING THE GROUNDWORK FOR GROWTH

Remember all that work we did getting your greenhouse foundation just right? Well, your plants need their own foundation too. And just like you wouldn't build your greenhouse on a pile of marshmallows (tasty as that might sound), your plants aren't too keen on growing in subpar soil.

Soil is your plants' entire world! It's their home. It's where they'll stretch their roots, grab their nutrients, and hopefully live their best plant lives. So, while we spent the last chapter making sure your greenhouse wouldn't topple over, now it's time to ensure your plants have an equally solid footing.

Let's get the dirt on dirt!

UNDERSTANDING SOIL: IT'S NOT JUST DIRT, IT'S DINNER!

Ever wonder why certain plants flourish while others languish in the very same location? Or have you ever gotten to peek into your neighbor's garden and wondered why their tomatoes look like they're auditioning for a magazine while yours resemble something from a vegetable horror film?

The answer may not be that fancy watering can or their alleged "green thumb"—the answer might be right under your nose … or feet, that is. It's all in the soil.

Let's break down this whole soil business into something digestible (pun absolutely intended). Your soil is basically your plants' all-you-can-eat buffet, complete with a drink station (water retention) and comfortable seating (proper structure) and hopefully good food (nutrients).

So what's really going on inside that lump of earthy goodness?

Gritty, Silky, or Sticky: *Your Soil Texture Personality Profile*

If you were to grab a handful of dirt from your backyard, another handful from the local dog park, and another from your neighbor's backyard (uh, probably don't just go do that without permission), each handful would have a different *composition*. They'd have different levels of nutrients, organic matter, and balances of texture types.

When I say texture types, I mean this: "Soil" is actually a complex blend of sand, silt, and clay. These are the main building blocks of your dirt.

Let's take a look at what we're working with here:

- **Sand:** The party animal of soil particles. Great at drainage, terrible at holding onto anything. Like that friend who can never keep a secret.
- **Clay:** The clingy one. Holds onto water and nutrients like they're going out of style. Sometimes a bit too well.
- **Silt:** The well-adjusted middle child. Not too loose, not too tight. Just right.

Clay particles are really tiny, sand particles are bigger, and silt is in the middle. These size differences explain why clay packs so tight and hard when wet, while sand lets everything slip right through—kind of like trying to hold water in a sieve versus a bowl.

The perfect mix? It's called loam, and it's the soil equivalent of hitting the jackpot. It's got just enough of everything to keep your plants happy, healthy, and throwing down roots like they mean it.

The golden ratio for loam is roughly 40% sand, 40% silt, and 20% clay—like a well-balanced recipe where every ingredient plays its part without trying to steal the show. Search online for a "soil texture chart" if you wanna really nerd out on this.

SOIL AND FERTILIZATION

"But JR, even if I know the perfect balance, how does that help me plant my plants?" Well, you'll know what to look for when getting soil.

"But JR, JR, how do I look for it?" Easy, partner! I'm on my way there!

The point I'm making here is you don't wanna just go scoop up dirt from your yard, or your neighbor's, or the local dog park, willy-nilly. You want to *test* the soil you're going to use, especially if you don't purchase it, but are opting to use your backyard dirt.

You can buy a soil test online or at local garden centers, which will give you beaucoups of helpful information about your dirt. It's like taking blood labs with a doctor to check how your levels are doing, and getting recommendations (you know, when your cholesterol tattles on you about all your fast food "cheat days"). If you're serious about growing, I recommend it, especially if you're planting anything in your native soil (meaning, anything from the ground under your feet).

However, if that sounds like a bit much (like running a background check on someone before going on a first date), then there are simpler DIY options. You can search online for "DIY soil tests" to get a lot of fun, informative ideas that make you feel like you're back in elementary school doing science projects.

Here's one of the most basic tests below.

The Soil Test: *Getting to Know Your Dirt*

Before you start playing matchmaker between your plants and their soil, you'll want to do a little detective work.

Here's a fun test that doesn't require a lab coat (though you can wear one if you're feeling fancy):

1. Gather a soil sample from various garden spots.
2. Grab a handful of your soil, and lightly dampen it (just spray, or drop droplets).
3. Give it a little squeeze (gentle, we're not stress-testing it).
4. Try to form it into a ball with one hand.

Notice what happens ... If it:

- Falls apart immediately → Too sandy (Time to add some organic matter!)
- Sticks together like clay sculpture → Too much clay (Let's loosen things up!)
- Forms a loose ball that crumbles when poked → Perfect! You've got yourself some lovely loam.

Amendments for Soil Structure

Once you know your soil's texture, you may feel your soil needs a makeover. This is where **soil amendments** come in. Amending soil means adding materials to improve its structure and fertility. Think of it as upgrading your soil's capabilities—like installing better hardware in a computer—by mixing in specific ingredients that address its weaknesses and enhance its strengths.

Soil amendments are your toolkit for transforming mediocre soil into an ideal growing medium. Each serves a specific purpose.

Organic matter is your new best friend, as it is perhaps the most valuable amendment of all. Organic matter is stuff like compost, aged manure, or leaf mold. It improves soil structure, feeds beneficial microorganisms, and slowly releases nutrients your plants need to thrive. If your soil has too much clay or too much sand, adding organic matter is a great all-around go-to.

Here are some other ideas:

Perlite and vermiculite improve soil structure in different ways.

- Perlite: Volcanic material that creates air pockets and improves drainage.
- Vermiculite: Mineral that retains moisture while maintaining good aeration.

For sustainable gardening, coconut coir offers an eco-friendly alternative to peat moss. Made from coconut husks, it provides similar benefits: excellent water retention and aeration.

Work in organic matter for clay soil challenges. Gypsum can also be effective. It breaks up compacted clay, improving both drainage and root penetration to create a more hospitable environment for plant growth.

For sandy soil challenges, work in organic matter. Mix in vermiculite or coconut coir for water retention. Layer on mulch to prevent moisture loss.

Other creative amendments exist out there for all you overachievers! Here are a couple of examples. Mycorrhizal fungi, often added as inoculants, are like tiny root extensions for your plants, forming underground networks that help them access more nutrients. Biochar, a form of charcoal, improves soil aeration and water retention, creating a hospitable environment for beneficial microbes.

Needless to say, there are lots of ways to turn soil from mediocre to good, and from good to great. It takes time to really build soil, but it's the gift that keeps giving.

A Word on Sterilization

While I'm generally all for letting nature do its thing, sometimes greenhouse soil needs sterilization to prevent disease. If you're buying from the store, you can skip this. But if you're reusing soil and want to get picky, you may want to look into this. You can:

- Use solarization (let the sun cook it, usually under a cover of some kind)
- Steam sterilize (for the serious folks)
- Buy pre-sterilized mix (no shame in the easy route)

pH: *The Mood Ring of Your Soil*

Remember mood rings? Well, soil pH is kind of like that, except it actually means something.

pH measures how acidic or alkaline something is. Sort of like a thermometer for sweetness to bitterness. The pH scale is from 0 to 14, and smack in the middle at 7, everything is neutral, as in pure water. Below 7, you're in acid territory (like vinegar or coffee), and above 7, you're in alkaline country (like soap or baking soda). The farther you move from 7 in either direction, the more intense it becomes, with 14 being the most alkaline, and 0 being the most acidic.

Most plants prefer their soil slightly acidic to neutral (6.0–7.0 pH), like a very mild vinaigrette rather than a lemon juice shot. Some plants like more acidic conditions (e.g. blueberries), and some like more alkaline conditions (e.g. asparagus), so be sure to check plant labels and ask local garden store workers and gardeners to find your specific plant's needs.

Testing pH is easy with a home kit & pH strips, which really look like a mood ring, giving different colors according to pH levels.

Once you know a soil's pH, adjusting it isn't rocket science, especially on a smaller scale like in a greenhouse:

- Too acidic? Add some lime (the mineral, not the fruit). Also, bone meal and oyster shells.
- Too alkaline? A bit of sulfur will bring that pH down faster than a dad joke at a family dinner. Also, pine needles/sawdust and granite rock dust have been thought to help.

There are other quick and dirty ways to run DIY tests on pH (think: baking soda and vinegar volcano vibes, from 4th grade science class). Just search for "DIY soil pH tests" online if you're eager.

Organic Matter: *The Life of the Party*

If soil were a party, organic matter would be the host that keeps everything running smoothly. It:

- Improves soil structure
- Feeds beneficial microbes
- Holds water (but not too much)
- Provides slow-release nutrients
- Makes your soil feel like a cozy blanket for roots

Think compost, well-rotted manure, leaf mold—basically, anything that used to be living and has broken down into dark, crumbly goodness.

Adding organic matter to soil is like giving your plants a multivitamin. It boosts fertility and nutrient levels, improves structure and reduces compaction, increases the soil's ability to hold water and nutrients, can help with disease, pests and weeds, and even detoxify polluted soil. It basically transforms your soil into a thriving ecosystem. I mean ... what *doesn't* it do?

Adding organic matter to soil may not solve *every* problem ... *but it sure feels like it does!*

So, as you bring your greenhouse problems to other gardeners, don't be surprised if you hear the advice repeated, "Have you tried adding some compost?" It sounds like a wonder drug, probably because it is.

You can buy compost and other organic material, or make piles of your own. We'll talk about that in a bit. To give your plants this gift, you can work it into the soil a little bit or just layer on top. I really like a layer of compost, followed by a layer of mulch (wood chips, straw, etc.).

Shopping for Soil: *A Buyer's Guide to Garden Center Glory*

Ever stood in the garden center, staring at walls of different soil bags like you're trying to decode ancient hieroglyphics? Been there! Let's break down these mysterious bags of dirt into something that makes sense.

Potting soil

Your container plants' best friend. It's a carefully crafted mix of materials like soil, peat moss (or coco coir for the eco-conscious), perlite, vermiculite, and composted materials. Think of it as the memory foam mattress of the plant world—light, fluffy, and perfect for roots to stretch out in.

Ideal for:

- Container plants
- Young transplants
- Anything in pots!

Garden soil / planting mix

The heavyweight of the soil world. This is actual soil that's been enhanced with organic matter and nutrients. It's denser than potting soil and great for larger raised beds or mixing into your existing greenhouse soil.

But careful! Not for container use. Using it straight in containers can lead to soil compaction like an elephant jumped on it.

For smaller to average-sized raised beds, some gardeners like a 50/50 mix of garden soil and potting soil. Some soil suppliers even have a "raised bed mix" which is probably a similar composition.

Seed starting mix

The baby blanket of soils—extra fine and sterile for those precious, tender seedlings. Usually, it actually doesn't contain soil (which is great for starting seeds, believe it or not). If potting soil is memory foam, this is like a cloud made specifically for plant babies.

Specialty mixes

Like ordering a custom drink at your favorite coffee shop, these are tailored for specific plants:

- Cactus/Succulent Mix (extra drainage for your desert dwellers)
- African Violet Mix (when your fancy plants demand the fancy stuff)
- Orchid Mix (basically bark chips playing dress-up as soil)

Pro tips for Soil Shopping

1. Check the bag's moisture content—if it feels like a water balloon, move along.
2. Look for chunks of varying sizes (uniform particle size often means lower quality).
3. Give it a sniff—good soil smells earthy, not sour or moldy.
4. Avoid bags with mushrooms growing out of them. (Yes, I've seen it happen!)
5. Check the expiration date—yes, soil can expire, usually after about a year.

The Budget-Savvy Approach

Buy basic potting soil and soup it up yourself with:

- Extra perlite for drainage
- Worm castings for nutrients
- Compost for organic matter & nutrients

Remember: sometimes the premium stuff is worth it, but you can usually enhance the basic options with a little DIY magic!

While we're adding extra sparkle to your soil, it's time to mention:

FERTILIZERS: THE PROTEIN SHAKES OF THE PLANT WORLD

Now, let's talk about giving your plants a boost. Your plants and soil need nutrients, which they would normally get in the wild. In a controlled environment, supplementation is often needed.

When it comes to fertilizing, you've got two main camps:

Synthetic fertilizers

Think of these like energy drinks for your plants. In a pinch, they can help with quick nutrient availability, to jolt your plants awake.

They don't really feed your soil like organic fertilizers do, however. For the long game, organic is the way to go.

Here's the scoop on synthetic fertilizers:

- Fast-acting
- Precise nutrient ratios
- Usually cheaper
- But! Can burn plants if overapplied
- Don't improve soil structure
- Can negatively impact environment, and create toxic buildup in soil

Organic fertilizers

These are more slow-release nutrition for your plants, so patience is key.

Unlike synthetic fertilizers that act like energy drinks, these natural options are more like a slow-cooked meal: they take time to break down, but they'll feed your plants better in the long run.

Organic fertilizer feeds the *soil*, not just the *plant*. It literally gets to the root.

Organic fertilizers:

- Release nutrients gradually
- Improve soil structure
- Feed beneficial microbes
- Environmentally friendly
- Less chance of over-fertilizing

Examples:

- Compost (the all-purpose superhero)
- Well-rotted manure (nature's original plant food)
- Various organic amendments (see below)

Now, before we dive into organic soil amendments, let's talk about your soil's nutritional needs. Just like us humans, plants need a balanced diet to thrive.

Soil Amendments for Boosting Nutrients

The heavy hitters in the plant nutrition world are what we call **macronutrients**: nitrogen, phosphorus, and potassium. These three are the star players you'll find listed as N-P-K on fertilizer bags (think of them as your soil's nutritional facts label).

Believe me, getting these right makes the difference between plants that merely survive and ones that throw a full-on growth party in your greenhouse.

Depending on what nutrients need boosting (you can find this out on a soil test), here are some examples:

- **Nitrogen (N):** aged manure, seaweed, fish, blood meal, alfalfa meal, feather meal, fish meal or emulsion, mushroom compost, or rice hulls.
- **Phosphorus (P):** bone meal, rock phosphate, fish emulsion, animal manure, lime, and seabird guano (cuz I know you have some of that lying around).
- **Potassium (K):** wood ashes, seaweed, dolomite lime, oyster-shell lime, kelp, gypsum, rock dust, or greensand.

SOIL AND FERTILIZATION

You might test the soil and see there are a lot more things on there than just N-P-K. You can also boost other nutrients in your soil like calcium (Ca) (eggshells!), magnesium (Mg) (lime!), sulfur (S) (elemental sulfur!), plus other minerals and nutrients.

Just take your nutrient deficiencies to the world wide web! Search online for "organic amendments to boost _____" and fill in the blank with the desired nutrient. The internet is a wonderful thing.

Once you choose your fertilizer options, you'll probably have a few more questions, such as how much, when, and how to apply it. Let's address those questions.

Fertilizer Amounts

The amount you'll need depends on three key factors: your soil's current nutrient levels (another reason to get that nifty soil test!), what you're growing, and which fertilizer you're using.

For organic fertilizers, here's a cheat sheet:

- **Light feeders (herbs, beans):** About 1–2 tablespoons per sq ft.
- **Medium feeders (peppers, cucumbers):** 2–3 tablespoons per sq ft.
- **Heavy feeders (tomatoes, squash):** 3–4 tablespoons per sq ft.

But here's the catch—these are starting points, not commandments. Always check the instructions for your specific fertilizer, as concentrations vary widely between products. Concentrated bat guano will pack more punch than general composted manure.

Pro tip: Start with less than you think you need. You can always add more, but removing excess fertilizer is like trying to un-salt your soup—technically possible but a real hassle. Monitor your plants' response and adjust accordingly. Yellow leaves? Add a bit more. Burning leaf edges? Dial it back. Your plants will tell you if you're getting it right.

Research your plant's requirements, your soil's current needs, and your fertilizer's instructions. Then go pick another gardener's brain. You'll get the hang of it.

Surface vs. Subsurface Application

Two basic ways to apply fertilizer are: surface application and subsurface application.

Think of surface application as spreading butter on toast—you're laying nutrients right on top of the soil. This method is great for covering large areas and slow-release fertilizers, but it takes longer to reach the roots. Some nutrients might get lost to the wind or wash away before plants can use them.

Subsurface application is more like stuffing a sandwich—you're getting nutrients down where the roots can find them. Dig a small trench, add fertilizer, and cover it back up. It takes more time, but it delivers nutrients right to your plants' doorstep, and you'll lose less to the environment. This method is perfect for intensive greenhouse growing, where every nutrient counts.

Surface application

- Pros:
    - Quick and easy to apply (like butter on toast)
    - Great for covering large growing areas at once
    - Perfect for pre-planting soil preparation
    - Easier to measure and spread evenly
    - Works well with granular organic fertilizers
- Cons:
    - Can increase greenhouse humidity
    - Nutrients may form surface salt crusts
    - Some loss of nutrients from runoff, before plants can access them
    - Takes longer to reach root zones
    - Can attract pests to the soil surface in an enclosed space

Subsurface application

- Pros:
    - Delivers nutrients directly to the root zone
    - Reduces salt buildup on soil surface
    - Better nutrient retention
    - Lower impact on greenhouse humidity

- More efficient in a controlled greenhouse environment
- Cons:
    - More time-consuming to apply
    - Requires careful placement to avoid root damage
    - Can be tricky with closely spaced plants
    - More physical effort required
    - Harder to achieve even distribution

Application Methods: Continued

Side-dressing (which has nothing to do with salad, by the way) means tucking nutrients next to your growing plants. Side-dressing can be done either on the surface or slightly below, depending on the type of fertilizer and the needs of your plants.

To side-dress under the surface, apply fertilizer in a shallow trench or band around the plant. In large planting areas like a bed or raised bed, a great way to accomplish this is to make a strip down the side of each row of plants. Then, cover back with soil, and water thoroughly.

With synthetic fertilizers, you'll want to apply at least 6 inches away from the main plant stem to prevent damage.

When side-dressing with organic fertilizers, such as compost, manure, or other natural amendments, the distance from the stem matters less than with synthetic fertilizers. You can safely apply them closer to the stem, typically within 2–3 inches, especially for established plants. With younger plants, you may want to add an inch, since they may be sensitive to nutrient surges.

And then there is **foliar feeding.** Think of foliar feeding as giving your plants a nutrient spa treatment—instead of waiting for nutrients to work their way up from the roots, you're spraying a light mist of liquid fertilizer directly onto the leaves, where they can absorb it through their pores (like a leafy face mask, if you will).

Foliar feeding works best for: quick nutrient fixes (think of it as plant first aid), and micronutrient deficiencies.

To do it: Mix a water-soluble fertilizer to half-strength, add a drop of mild dish soap as a spreader (helps the solution stick to leaves instead of beading up like morning dew on a lotus leaf), and spray the undersides of leaves early morning or late afternoon when it's cool. Your plants will be strutting their stuff with that post-spa glow in no time!

Strategy: Dilute more than package directions suggest for greenhouse use—enclosed spaces mean more humidity and slower drying time.

Timing Your Fertilizer Applications

Think of fertilizing like feeding teenagers—timing and frequency matter, and everyone's hungry at slightly different times. In a greenhouse setting, your feeding schedule needs to match both the season and your plants' growth stages.

Seasonal timing:

- **Early spring** *(every 4 weeks)***:** Start with a balanced feeding as plants wake up.
- **Mid-spring through summer** *(every 2 weeks)***:** Peak feeding season, especially for heavy feeders.
- **Late summer** *(every 4 weeks)***:** Begin tapering off as growth slows down.
- **Winter** *(minimal/every 6 weeks)***:** Light feeding every six weeks or skipping entirely for dormant plants keeps things low-maintenance. The exception is if you are growing year-round crops.

Specific timing according to growth stage:

- **Seedlings:** Hold off until first true leaves appear.
- **Vegetative growth:** Regular feeding every 4–6 weeks.
- **Flowering/fruiting:** Increase frequency to every 3–4 weeks.
- **End of season:** Reduce or stop to avoid forcing new growth.

True leaves, by the way, are the first real leaves a seedling produces after its initial **cotyledons** (seed leaves). These are typically more complex in shape and resemble the mature plant's foliage.

Pro tip: With organic fertilizers, apply a bit earlier than you think you need to—they're like crockpot cooking, taking time to break down and become available to your plants. For example, if you want nutrients available for spring growth, work them into your soil about a month before.

SOIL AND FERTILIZATION

Applying fertilizer in the **early morning** is best, or **late afternoon** is second best. This avoids the heat of midday, reducing the risk of leaf burn if fertilizer splashes and ensuring plants can absorb nutrients efficiently without stress.

Remember: Your greenhouse creates a warmer environment where organic matter breaks down faster than outdoors. This means nutrients cycle through more quickly, but it also means you might need to feed a bit more frequently than in an outdoor garden. Let your plants' growth and leaf color be your guide—they'll tell you if they're hungry for more.

Now, if all the above seems a bit confusing and/or overwhelming, a *really* simple version is to feed every month. Hold off with the seedlings til you see those true leaves—it's like giving steak to a baby. Think of this as a monthly check-in: Feed when the calendar flips, and you're good to grow!

Final Fertilizer Thoughts

Greenhouse fertilizing with organic materials is like slow cooking—it's all about patience and layering those nutrients. The enclosed environment means your organic matter breaks down differently than outdoors, so start with about half the recommended amounts and let those nutrients build up naturally.

Keep your fertilizer storage as organized as a fussy librarian—label everything clearly, store in airtight containers in a cool, dry spot, and away from direct sunlight. For dry ingredients, treat moisture like your fertilizer's archnemesis. Think of it like keeping your breakfast cereal fresh and crunchy—nobody likes that stuff soggy!

A well-organized fertilizer shelf means you'll always have the right nutrients ready when your plants need a boost. Use clear labels that even a sleepy gardener couldn't mix up. There's nothing worse than playing "guess which decomposed matter this is" at 7 AM.

Think your sickly-looking plant needs medicine? Before reaching for the fertilizer bag when a plant looks unwell, take a step back. It's essential to diagnose the underlying issue—be it pests, diseases, or environmental factors—before considering fertilization. Adding nutrients won't cure all ailments and might even make things worse if misapplied.

The secret to success with organic fertilizing in a greenhouse? Think long-term and keep observing. These natural nutrients release slowly, building a robust soil ecosystem over time. Watch your plants' response, maintain consistent feeding schedules, and remember—more than simply feeding plants, you're nurturing an entire soil food web.

With synthetic fertilizers, think observation and moderation. Watch your plants, start gentle, and remember—it's easier to add nutrients than to deal with fertilizer burnout.

Your plants will tell you what they need; you just need to give them time to adjust, and learn to speak their language.

MULCH: YOUR SOIL'S FAVORITE BLANKET

Let's talk about tucking your soil in at night—or as fancy gardeners call it, **mulching**! Think of mulch as your soil's personal comfort system: a cozy blanket in winter, a cooling shade umbrella in summer, and a bouncer keeping those pesky weeds from crashing the party.

Why mulch? Well, why do you wear a hat on a sunny day? Same idea! Mulch:

- Keeps moisture in (like a lid on your coffee cup)
- Regulates soil temperature (think thermostat for your dirt)
- Blocks weeds (nature's "do not disturb" sign)
- Slowly feeds your soil as it breaks down (the gift that keeps on giving)

Mulch Options (From Fancy to Free):

- **Straw:** a classic choice—just make sure it's seed-free, unless you're going for that "surprise grass everywhere" look.
- **Wood chips:** long-lasting, but avoid fresh ones right next to plants—they're nitrogen hogs.
- **Leaves:** nature's free mulch delivery service.
- **Grass clippings:** only if they're herbicide-free—we're feeding plants, not poisoning them.
- **Compost:** a double-whammy, it protects AND feeds.

Pro tip: In the greenhouse, keep mulch a bit lighter than outdoors. You're already controlling the climate, so you don't need the heavy-duty stuff. Think light cardigan rather than winter coat.

The Golden Rules of Mulching

1. Don't pile mulch against plant stems (nobody likes a wet collar).
2. Apply 2–3 inches deep (like porridge—not too much, not too little).
3. Refresh as needed (when it starts looking like it needs a makeover).

At minimum, adding compost annually is a great way to maintain soil health. Your greenhouse soil is like a bank account—make regular deposits of organic matter and nutrients, and your plants will enjoy rich returns!

COMPOSTING: WHERE KITCHEN SCRAPS GO TO GRADUATE

Remember that feeling when you transformed your leftovers into a gourmet meal? Well, composting is like running a transformation academy for your kitchen scraps and garden trimmings. Instead of ending up in a landfill having an existential crisis, these organic materials get to fulfill their destiny as black gold for your garden.

The Magic Behind the Mess

Here's what's really happening in that pile of potential: tiny microbes (think of them as your personal decomposition team) are having an all-you-can-eat buffet. These microscopic munchers break everything down into rich, crumbly compost that makes your plants do a happy dance. It's like they're turning yesterday's salad into tomorrow's garden feast!

Greenhouse Composting: Small Space, Big Impact

You can compost in a bin, you can compost in a pile. All it needs is air and water, and your scraps will make you smile.

Okay, so poetry's not my strong suit. But compost is pretty simple.

Really, all compost needs is 4 things: greens, browns, air, and water. Here's that same recipe again, but more detailed: If you have a balanced mix of …

1. "green materials" (nitrogen-rich materials) and
2. "brown materials" (carbon-rich materials),
3. *and* you keep your compost aerated by turning it,
4. *and* you keep it moist …

You'll be a garden alchemist, turning junk to black gold, in no time! (Well compost does take some time, actually, but it's worth the wait!)

All you need is a spot to start piling on the treasure. You can compost in the open air in: a plastic trash can, a wire cage stabilized by four posts, a corral made from pallets, composting bays made from corrugated metal and wood (look for DIY plans online), or a simple pile in the corner.

"But JR," you might say, "my greenhouse barely has room for my tomato dreams, let alone a compost pile!"

Fear not, my space-conscious friends! Enter the world of compact composting:

There are **compost tumblers:** sealed bins you can spin to aerate, speeding up decomposition in a tidy, contained system.

Also, there are **countertop composters:** electric or manual units that dehydrate and grind scraps into compost quickly.

You could also try **vermicomposting**, where red wiggler worms in a container break down food scraps into nutrient-rich compost (like having tiny composting machines).

Or there's **bokashi composting**, which is typically done in a specific airtight bucket you can place under a sink or table. This is a fermentation method that uses beneficial microbes to break down even meat and dairy (think pickling your scraps without the odor).

These are compact, efficient, and ideal for greenhouse corners where a traditional compost pile isn't practical.

The Recipe for Success

Like any good recipe, composting needs the right ingredients in the right proportions:

- **Greens (nitrogen-rich):**
    - Fresh grass clippings (lawn haircuts)
    - Vegetable scraps (your cooking prep leftovers)
    - Coffee grounds (yes, your plants love caffeine too!)
    - Fresh plant trimmings

- **Browns (carbon-rich):**
  - Dried leaves & twigs (autumn's gift to gardeners)
  - Straw or hay (not the fresh stuff)
  - Shredded paper and cardboard (finally, a use for those Amazon boxes)
  - Wood chips (think small chips and sawdust, not lumber—untreated wood)

Mix these in roughly equal parts—think of it as making a layer cake, but instead of cake and frosting, you're layering browns and greens. Some people like to add more browns than green (2:1 or 3:2, for instance). Like chili, everyone has their own compost recipe they swear by.

Add more greens (e.g., food scraps, grass clippings) if the pile is dry, slow to break down, or not heating up. Add more browns (e.g., leaves, cardboard) if the pile smells bad or is too wet and slimy.

Some things don't go so well in compost, like dairy, meat, oils, plastic, synthetic stuff, excrement from carnivores, and weeds gone to seed.

To find out more extensive lists of greens/browns, as well as what to compost and not to compost, you can search online for "compost chart" to get something printable to put up by your compost.

Moisture & Air: The Other Two Big Ones

Your compost should feel like a wrung-out sponge—damp enough to keep the microbes happy, but not so wet they need swimming lessons.

And don't forget to turn it occasionally! Think of it as letting your compost pile breathe—everyone needs a little fresh air now and then.

Where to Put Your Pile

In or around your greenhouse, pick a spot that's:

- Easy to reach (no Olympic hurdles to add kitchen scraps)
- Out of direct sunlight (think at least partial shade—your compost doesn't need a tan)
- Away from plant roots (they'll get the goods when they're ready)
- Close to a water source (for those times it needs a drink)

The "Is It Ready Yet?" Test

Finished compost looks like rich chocolate cake crumbs and smells like a forest floor after rain. It should be dark, crumbly, and earthy-smelling, with no recognizable food scraps or large pieces of organic material left. If it still looks like last week's salad, give it more time. You can't rush perfection!

If it looks and smells like rich soil, then it's good to go!

If you're really on your maintenance, compost can take 2 to 6 months to mature. If you're more hands-off with it, it may take 6 months to a year. Regular turning (every 1–2 weeks) speeds up the process.

Using Your Black Gold

Once your compost graduates from the transformation academy, it's ready to work miracles in your garden:

- Top-dress existing plants (like an organic blanket of goodness)
- Mix into potting soil (supercharging your soil mix)

Remember: Composting isn't just recycling—it's upgrading! You're not only saving kitchen scraps from the landfill; you're creating premium plant food that would make any garden center jealous. You get to feel like you're helping the planet, pampering your plants, and turning scraps into treasure —all in one satisfying, earth-friendly swoop.

Not bad for a pile of what most people throw away, right?

CHECK-IN

Well, here we are again, and it's time for some well-deserved celebration. You dug deep. You got in your plants' world. And you emerged with some serious down-to-earth wisdom.

No longer will you mistake any old clump of earth for quality growing medium. By working through this chapter, you've gained some valuable know-how. You:

- Understand soil texture types and can identify if you're working with sandy, silty, clay, or that garden jackpot—loam.

SOIL AND FERTILIZATION

- Know how to perform a simple soil test to determine your soil's composition and needs. (No lab coat required!)
- Can choose appropriate amendments to transform mediocre soil into plant paradise
- Recognize the importance of pH and how to adjust it for different plant needs.
- Understand the difference between synthetic and organic fertilizers, and how to apply them properly.
- Have learned the magic of mulching and composting—turning "waste" into gardening gold.

Let's check where you stand with your soil:

- Have you determined what type of soil you'll use in your greenhouse—purchased potting mix or amended garden soil?
- Do you understand which amendments might benefit your specific growing conditions?
- Have you considered your plants' nutritional needs and selected appropriate fertilizers?
- Do you have a plan for building soil health over time through composting or regular organic matter additions?

Don't worry if some points still feel fuzzy—that's normal. The key is that you're making progress. Which of these skills makes you feel most confident? That's your foundation to build on!

Here are some concrete steps you can take, if you haven't already:

- Gather soil samples from your growing area (or purchase quality potting mix) and perform a simple texture test.
- Start a compost collection system, even if it's just a small bin or tumbler near your greenhouse.

Quick success tip: The time you invest in soil preparation now will pay dividends all season long. Think of it as your greenhouse's foundation beneath the foundation—plants with happy roots grow vigorously, resist pests and diseases better, and produce more abundantly. A few extra hours spent on soil now means fewer problems and more harvests later!

Wrap-Up

Look, I know we've covered a lot of ground here (pun intended!). Maybe your head's spinning, but it's also grown! (Well, metaphorically at least). Remember when soil testing seemed complicated? Now you're practically a soil scientist!

If you're looking for a *very* detailed look at soil, composting, and all things organic gardening, I'd highly recommend a book from my fellow author, Josie Beckham: *Permaculture Gardening for the Absolute Beginner*. It's loaded with wisdom and Josie's awesome personality, and you'll come out ready to garden (and maybe even have some chickens).

Remember: Whether you're working with premium potting mix or amending your backyard dirt, the key is understanding what your plants need and creating a living soil system that supports them. It's not about perfection; it's about progress. Your soil will improve over time, just like your gardening skills. Keep learning, keep growing, and most importantly, keep having fun with it!

And hey, now that we've got your soil situation sorted, you're probably itching to get some plants in there. Well, good news! That's exactly what we'll cover in our next chapter. Get ready to roll up those sleeves and start playing plant matchmaker—we're about to populate your greenhouse with some green friends!

8

PLANTING YOUR FIRST PLANTS

CHOOSING, PLANTING, AND GROWING WITH SOLID STRATEGY

This is it! The moment you've been waiting for! Your greenhouse's opening night.

After all that prep work, it's finally time for the star performers to take the stage. Like a Broadway director on opening night, you're about to fill that beautiful space with actual living talent. Because let's face it, what's a theatre without its actors? (Or in our case, what's a "greenhouse" without its "green?")

But you may wonder: *"Where do I even begin?"*

Maybe you're standing in your brand-new greenhouse, trowel and watering can in hand, feeling like a kid who just got the keys to a candy store but doesn't know which jar to open first. The air is rich with possibility (and maybe a hint of anxiety). Your heart's doing a little happy dance while your brain's throwing around questions faster than a squirrel with too much coffee.

"What if everything dies? What if I'm actually terrible at this? Why didn't these seeds come with GPS navigation?"

First, take a deep breath. That slightly panicky feeling? Totally normal. In fact, I'd be a bit worried if you weren't at least a little nervous about nurturing these tiny life forms.

Here's the thing: plants actually want to grow. They've been doing it for many, many years without human help. Think of yourself less as their master gardener and more like their enthusiastic cheerleader with a watering can.

GETTING TO KNOW YOUR PLANTS

Easy-Win Plants: Your Greenhouse Training Wheels

Remember learning to ride a bike? You probably didn't start with a double backflip. Same goes for greenhouse gardening.

Let's start with some plants that are basically the golden retrievers of the plant world—friendly, forgiving, and eager to please.

Leafy greens

- Lettuce
- Spinach
- Kale
- Arugula
- Swiss Chard

Root veggies

- Radishes
- Carrots
- Beets
- Garlic
- Onion

Other veggies

- Cucumbers
- Squash
- Hot & bell peppers
- Tomatoes
- Legumes (green beans, peas)

Hardy herbs

- Basil
- Mint (Warning: Grow only in a container! Will take over if given the chance—think teenager in the bathroom)
- Parsley
- Cilantro
- Perennials like mint, rosemary, thyme, and sage come back every year.

A few confidence boosters

- Microgreens (ready in days—perfect for the impatient gardener)
- Cherry tomatoes (like regular tomatoes but with less pressure)
- Snap peas (they literally tell you when they're ready by getting plump)

Still wondering what to plant first? A quick trip to your garden center with your notebook will yield great results. Just tell the workers, "Hey! I've got a new greenhouse. What plants do well in this area?" And let the enthusiastic support and wisdom roll in ...

Also, here are some good questions to ask yourself:

- How much space do I have in my greenhouse, and what are my climate conditions? *(Sunlight, plus accounting for any of your greenhouse climate control adjustments.)*
- What plants do my family and I enjoy eating or using the most?
- Do I want quick results or long-term crops? *(A combination of both can be the most rewarding.)*
- What are the easiest plants to grow for beginners?
- Can I add some perennial plants that will come back year after year?
- Are there native plants that would thrive in my greenhouse? *(Ask local gardeners for tips!)*
- What are some expensive foods I'd love to grow myself to save money?
- Do any of the plants I want fall into the "dirty dozen" (those most affected by pesticides)? *(Think strawberries, kale, tomatoes.)*

- Can these plants grow year-round in my greenhouse, or will I need seasonal adjustments?
- Do I have the time and energy to care for these plants?
- Could I share, swap, or even sell extra seedlings or produce with others? *(Great for building community!)*

And if you really want beginner-friendly, here are some choices from our training wheels list that are especially kind to young gardeners:

Lettuce, kale, spinach, arugula, and swiss chard:

- Why? These leafy greens thrive in a wide range of temperatures and grow quickly. They don't require extensive care or perfect conditions and are forgiving of rookie mistakes like inconsistent watering. They also provide quick rewards, which keeps motivation high.

Legumes (green beans and peas):

- Why? These plants are not fussy about soil conditions and can grow well with minimal input. Peas, in particular, can handle cooler temperatures, while green beans are fast growers that provide clear signs of progress. Their climbing nature is perfect for vertical space use, making them greenhouse-friendly.

Radishes:

- Why? Root vegetables like these are hardy and require little more than well-drained soil and consistent watering. Minimal pest problems and require little care. Radishes grow incredibly fast, often yielding results in under a month (from seed to snack in about 3 weeks!), making them ideal for beginners eager for quick wins.

Herbs (Hardy Varieties Like Basil, Parsley, and Mint):

- Why? Hardy herbs are resilient and require minimal attention. They grow quickly in a greenhouse environment and can thrive even with uneven care. Plus, they're small and space-efficient.

Pro tip: Start small! Better to succeed with three plants than get overwhelmed by thirty. You can always expand your plant family later when you're feeling more confident.

Growing Regions: Nature's Climate Club Memberships

Ever wonder why your cousin in Minnesota can't grow the same plants as your aunt in Florida? It's because they belong to different "growing regions"—think of them like nature's climate clubs, each with their own dress code and party rules.

The USDA (United States Department of Agriculture) has divided North America into zones based on average minimum winter temperatures, ranging from Zone 1 (the "don't forget your parka" zone) to Zone 13 (the "eternal summer" zone). These zones help gardeners figure out which plants will survive and thrive in their area, kind of like a matchmaking service for plants and places!

But here's where greenhouses get interesting—they're like having a VIP pass that lets you bend these rules a bit. While you can't completely ignore your growing zone (it takes some serious chops to mimic a tropical rainforest in Alaska!), a standard greenhouse can usually bump you up by 1–2 growing zones, giving you a longer growing season and more plant options. Think of it as nature's cheat code!

However, your base zone still matters because it affects things like how much heating or cooling you'll need, what kind of insulation to use, and how to handle seasonal changes. That's why checking your growing zone (just search "USDA growing zone + your zip code") is like getting the inside scoop on what nature's giving you to work with ... Because what works in Maine might make your plants melt in Arizona!

On that note, let me clear up the greenhouse glass on something—this book is written with a Northern Hemisphere garden in mind. More specifically, I've been writing from a North American perspective because that's home territory for me.

But before my overseas gardening buddies start feeling left out, let me say this loud and clear: You're absolutely included in this party! Can you feel the love I'm sending over there? The core principles here work anywhere; you just might need to tweak a few details, like sun patterns, growing regions, and maybe the timing of your plantings.

Think of it like borrowing a recipe from a friend—adjust the spices to your taste, but the dish still comes out delicious. Whether you're in sunny Sydney, bustling Berlin, or cozying up in Calgary, the tips in this book can help you create your perfect greenhouse oasis.

Understanding Plant Personalities (Yes, They Have Them)

Just like people, plants have their own quirks and preferences. Some are morning glory people, perking up with the sunrise, while others are night owls, preferring cooler evening temperatures. Understanding these personalities helps you create their ideal living conditions.

Some like it hot, some like it cold. Or, rather, some like full sun and some like shade, and some like it in the middle.

Here's a concise breakdown of plant sunlight needs:

Full Sun: 6+ hours direct sunlight daily

- Think tomatoes, peppers, squash, most flowering plants
- Best for plants that fruit or flower

Partial Sun/Shade: 4–6 hours direct sunlight daily

- Think leafy greens
- Morning sun with afternoon shade often ideal

Light/Dappled Shade: 2–4 hours direct sun or filtered light

- Think ferns, hostas, some more delicate herbs
- Does well under shade cloth or tree canopy

Full Shade: Less than 2 hours direct sun

- Think peace lilies, moss
- Thrives in indirect light

Remember: Morning sun is gentler than afternoon sun, and greenhouses filter 10–20% of natural light, so adjust accordingly.

That's really all you need to know for basic light planning. When in doubt, check the plant's specific needs. You can find that online or on the back of seed packets.

In fact, seed packets have a lot of useful information about a plant's specific needs. You can also search online to find out things like: sunlight needs, watering needs, spacing requirements, soil and nutrient needs, and environmental conditions (temperature/humidity).

Some plants may have special considerations, like needing support. And I don't mean speaking positive affirmations over them every morning, although, hey, you go for it! I mean, like trellises or cages for tomatoes to grow up, or vertical supports for vining plants like cucumbers, pole beans, peas, or vining squash.

TIMING IS EVERYTHING (BUT DON'T STRESS TOO MUCH ABOUT IT)

Plants have their preferred seasons, and aligning your planting schedule with these natural rhythms ensures a continuous supply of fresh produce.

Cool season crops are like those friends who thrive in sweater weather—they grow best between 40–75°F, and *most* actually do fine in a light frost now and then (depending on the specific crop). Think leafy greens, peas, and root vegetables that flourish in early spring or fall.

Warm season crops, on the other hand, are your sun-loving summer crew—tomatoes, peppers, melons, squash, and cucumbers that live for temperatures between 65–85°F. These sun lovers won't tolerate frost (can you blame them?) and need warm soil to even think about sprouting. Once the temperature drops, they slow down like molasses in January!

One of the beauties of greenhouse gardening is that you can bend the rules a bit when it comes to growing seasons. That said, working with nature rather than against it makes life easier.

Search online for a "planting calendar" to see a full chart of what to plant when.

Seasonal Planting: Planning for Year-Round Harvests

With that knowledge, let's talk about turning your greenhouse into a year-round produce factory. While outdoor gardens take winter vacations, your greenhouse can keep cranking out harvests through every season. The trick? Strategic timing and picking the right plants for each season's conditions.

Understanding Growing Seasons

Think of your greenhouse as having four distinct growing seasons, each with its own star performers:

- **Spring (March–May)**
    - Perfect for: Peas, carrots, lettuce, radishes, beets, most herbs
    - Ideal temp range: 60–70°F
    - Key consideration: Protection from late frosts
    - Pro tip: Start summer crops like tomatoes indoors now.

- **Summer (June–August)**
    - Perfect for: Tomatoes, peppers, cucumbers, eggplants, basil
    - Ideal temp range: 70–85°F
    - Key consideration: Ventilation and shade management
    - Pro tip: Start fall crops in July while summer crops are still producing.

- **Fall (September–November)**
    - Perfect for: Broccoli, cauliflower, root vegetables, spinach, kale
    - Ideal temp range: 60–75°F
    - Key consideration: Maximizing light as days shorten
    - Pro tip: Cold-hardy crops planted now can produce through winter.

- **Winter (December–February)**
    - Perfect for: Spinach, kale, cold-hardy lettuces, hardy herbs, microgreens, root veggies
    - Ideal temp range: 50–65°F
    - Key consideration: Supplemental lighting and insulation
    - Pro tip: Focus on compact, quick-growing crops.

Creating Your Planting Calendar

Here's how to build a planting schedule that keeps your greenhouse productive year-round:

1. **Map your climate window**
   - Note your first and last frost dates.
   - Track typical temperature patterns.
   - Document light levels through seasons.
2. **Plan succession planting**
   - Plant new crops every 2–3 weeks.
   - Calculate maturity dates to prevent gaps.
   - Replace finished crops immediately.
3. **Match plants to conditions**
   - Choose varieties bred for greenhouse growing.
   - Group plants with similar temperature needs.
   - Consider vertical space utilization.

Pro tip: Create a simple spreadsheet tracking: planting dates, expected harvest times, and space requirements. Your future self will thank you when trying to remember exactly when you planted those carrots.

Extending Growing Seasons

Want to push those seasonal boundaries? Here's how:

- **Temperature management:**
  - Use row covers for extra warmth (+5–10°F).
  - Install thermal mass (water barrels, stone paths).
  - Consider heating mats for seed-starting.
- **Light enhancement:**
  - Add grow lights during short days.
  - Use reflective materials strategically.
  - Keep glass clean to maximize natural light.
- **Protection strategies:**
  - Cold frames within the greenhouse.
  - Floating row covers for frost protection.
  - Bubble wrap insulation on walls.

Remember: A greenhouse doesn't eliminate seasons—it just gives you more control over them. Work with nature's rhythms while taking advantage of your greenhouse's climate-controlling superpowers.

Keep notes on what works in your specific setup. Every greenhouse has its own microclimate quirks, and documenting your successes (and those "learning opportunities") helps refine your timing for future seasons.

THE PLANTING PROCESS

So you've got your space ready, you've chosen your soil, and you've picked out which plants you want to grow. It's time to plant some plants! *Now what?*

You may be wondering if you should plant seeds directly in the ground, or if you should buy some of those little seedlings from the garden center.

The classic "to seed or not to seed" question! Let's break this down into a tale of two growing styles, shall we? Think of it like choosing between adopting a teenager or having a baby—both are rewarding, but they come with different challenges and joys.

Starting from Seed: The Baby Route

Starting from seed is like being there for your plant's entire life story, from their first awkward sprout to their full-grown glory. You get to witness every wobbly step and growth spurt! Plus, you'll have bragging rights when your tomatoes start producing ("I knew them when they were just a seed!").

Advantages:

- WAY more variety (seed catalogs are like Netflix for plant nerds)
- Usually cheaper (like, way cheaper—we're talking pennies per plant)
- The ultimate smug gardening satisfaction ("Oh these? I grew them from seed.")
- Better control over growing conditions from day one
- No chance of bringing home hitchhiking pests from the garden center

Challenges:

- Requires more patience (Rome wasn't built in a day, and your pepper plant won't be either)
- Needs more equipment (lights, seed starting mix, trays)
- Higher learning curve (but hey, you like learning!)
- Some seeds can be finicky germinating (looking at you, lavender)

Starting from Seedlings: The Adoption Route

Seedlings are like getting a head start—someone else handles those delicate early days, and you can skip straight to the fun part! It's perfect for those "I should have started seeds two months ago, but here we are" moments.

Advantages:

- Instant gratification (because sometimes we just want plants NOW)
- Less equipment needed to get started
- Easier for beginners (no shame in a little bit of help!)
- Earlier harvests in your first season
- You can see exactly what you're getting

Challenges:

- More expensive per plant (convenience has a price tag)
- Limited variety (you're stuck with what the garden center offers)
- Possible transplant shock (imagine moving to a new house—plants feel that too)
- Risk of bringing home pests or diseases (always check those leaves!)

Here's a fun middle ground: why not try both? Start some easy seeds (hello, basil and zinnias!) while also picking up a few seedlings. That way, you can learn the seed-starting ropes without putting all your tomatoes in one basket. Plus, you'll get to experience both the "proud parent of a seed baby" and the "cool adoptive plant parent" vibes!

Ya know, I can still remember some of my first planting days. "Too deep? Not deep enough? Why are these seeds so small?" Second-guessing myself and probably planting way too many.

But you know what? The plants grew anyway. Some did well, some didn't, but each one taught me something. It's worth getting those early lessons under your belt, even if you stumble through it like I did. A garden's pretty forgiving, and trust me, those plants are rooting for you! (Terrible pun, J).

At the end of the day, whether you choose seeds or seedlings, you're still growing your own food and flowers, and that's pretty amazing. The plants don't care how they started—they just want to grow!

Some Tips to Not Kill Things (It's Easier Than You Think)

Let's run through some simple care tips that will prevent your plants from staging a wilting protest:

- **Watering wisdom:**
  - Check soil moisture before watering (stick your finger in—if it's dry past your first knuckle, time to water).
  - Water the soil, not the leaves (plants drink through their roots, not their faces).
  - Morning watering is best (gives plants time to dry before night).
  - *Refer to Chapter 5.*

- **Feeding your friends:**
  - Start with good soil (like building a house on sand vs. concrete).
  - Feed regularly during growing season (but don't overdo it—more isn't always better).
  - Watch for signs of nutrient deficiency.
  - *Refer to Chapter 7.*

- **Space management:**
  - Give plants room to grow (nobody likes a crowded elevator, including plants).
  - Consider mature sizes when planting (that cute little tomato seedling will become a beast).
  - Check space requirements on the back of seed packets and/or online.
  - Think vertical when possible (grow up when you can't grow out).

Remember to think about root depth as well. Different plants have specific root depth requirements, so research is key. For instance, tomatoes need deep planting to support their sturdy stalks, while lettuce is good with shallower depths.

Your First Planting Day: A Step-by-Step Guide

1. **Prep work:**
   - Gather your supplies (seeds, soil, containers, labels).
   - Make sure your soil is moist but not soggy.
   - Have your planting plan ready.
2. **Planting process:**
   - Follow seed packet depth guidelines.
   - Label everything (trust me, you won't remember what's what).
   - Water gently after planting.
3. **Initial care:**
   - Keep soil consistently moist until germination.
   - Watch for seedlings. (The most exciting part!)
   - Thin as needed (yes, it feels mean, but it's necessary).

Common Rookie Mistakes (And How to Avoid Them)

We all make mistakes. I definitely still do, and it's alright to make them. That said, here are some common ones to watch out for:

- **The Drowning:**
  - Overwatering is more common than underwatering
  - Solution: Check soil moisture before watering
- **The Overcrowd:**
  - Cramming too many plants together
  - Solution: Follow spacing guidelines (they exist for a reason)
- **The Neglect:**
  - Forgetting to check on plants
  - Solution: Make it part of your daily or regular routine
- **The Overfeed:**
  - Thinking more fertilizer = better
  - Solution: Follow feeding instructions and start light

When Things Don't Go As Planned (Because Sometimes They Won't)

Even experienced gardeners have plants that don't make it. It's not failure; it's *feedback*. Each "oops" moment is actually a learning opportunity in disguise. Keep notes on what worked and what didn't—this information is gold for your next planting adventure.

The Wait (And What to Do During It)

The time between planting and seeing those first green shoots can feel longer than waiting for water to boil. Here's how to stay busy (and sane):

- Start a garden journal (track what you planted and when).
- Take photos (before and after shots are amazing motivation).
- Plan your next plantings.
- Research recipes for your future harvest.

ADVANCED TECHNIQUES: LEVEL UP YOUR GREENHOUSE GAME

So you've mastered the basics, and now you're ready to take your greenhouse gardening from "pretty good" to "wait, are you secretly running a farmer's market?" Whether you're looking to maximize your harvests or streamline your planting schedule, these methods will help you grow smarter, not harder.

Companion Planting: Maximizing Space and Yield

Did you know some plants actually do better when paired with the right plant neighbors? This is known as the art of **companion planting.** Think of companion planting as nature's version of matchmaking—some plants just work better together, like a garden-variety buddy system! It's about strategically placing plants near each other.

Companion planting enhances plant growth, deters pests, improves pollination, optimizes space, enriches soil, reduces the need for chemical inputs, and promotes biodiversity for a healthier and more resilient garden. Now, how's that for teamwork?

Take the classic "Three Sisters" combo, for instance: corn, beans, and squash. The corn plays apartment building for the climbing beans, which return the favor by fixing nitrogen in the soil, while squash spreads out below like a living mulch, keeping weeds down and moisture in. It's like they formed their own little plant support group!

Or there's tomato and basil—great together in and out of the garden! This pairing isn't just about making your future margherita pizza more convenient. Like any great relationship, they help each other grow.

See, basil acts like your tomato's personal bodyguard, repelling pesky insects with its aromatic oils. Some say the basil actually makes your tomatoes taste better! Something about those fragrant compounds it releases helps enhance the tomato's flavor development. It's like basil is your tomato's life coach, helping it become its best, most delicious self.

Just plant a few basil plants around the base of each tomato (leaving enough room for everyone to breathe—we're going for cozy, not crowded!), and watch this dynamic duo thrive together.

But just like that aunt and uncle who can't be seated together at family dinners, some plants just don't play nice. Keep beans away from onions unless you enjoy garden drama. They bicker endlessly about lighting (sun needs), and beans get so offended by onions' strong personalities (their compounds) that it slows their growth.

There are a lot of plants out there, and a lot of options for pairing.

Want to dive deeper? Search for terms like "companion planting guide," "beneficial plant combinations," or "plants that grow well together." Also, if you have a few plants you know you want to grow, it makes searching easy. Take tomatoes, for instance. Just search "good companion plants for tomatoes," and you'll be matchmaking in no time.

Remember: good neighbors make good gardens, but like any relationship, it takes a bit of research to find the perfect match!

Let's Get Adventurous!

So you want to live a more exotic life, eh? Thinking outside the box of what grows locally, and dreaming of turning your greenhouse into a mini Bahamas? (Minus the tourist crowds and overpriced coconut drinks?) Well, grab your imaginary passport, because we're about to send your greenhouse on a tropical getaway!

Think beyond the humble tomato—I'm talking pineapples lounging like royalty and lemongrass swaying in your artificially humid breeze. These exotic showstoppers can transform your greenhouse from *"Oh, lovely herbs"* to *"Holy moly, is that a PINEAPPLE?!"*

Now, fair warning: exotic plants can be as demanding as a cat with a gourmet palate. They've got some specific needs. Bananas want their soil acidic, lemongrass needs its drainage on point, and don't get me started on the humidity requirements. These plants want it steamy like a tropical spa day, every day.

Your job is to recreate their dream home environment, minus the tropical storms and wandering monkeys. Set up humidifiers or misting systems (those tropical darlings love their moisture), and get ready to play the long game—these plants operate on island time when it comes to fruiting (a pineapple can take up to 2 years to reach maturity).

Remember: patience is your best friend here (well, that and a good humidity meter). But trust me, when you're sipping fresh-squeezed lime juice from your very own tree, you'll feel pretty awesome!

Succession Planting: The Netflix Binge of Gardening

Think of succession planting as the "continuous play" feature of your garden—instead of watching all your lettuce mature at once (and ending up with more salad than a rabbit convention), you plant small amounts every few weeks. This way, you get a steady stream of fresh produce rather than a sudden avalanche of vegetables demanding to be harvested *right now.*

Here's how to keep that show running: Start by planting a small section of quick-growing crops like radishes or lettuce. Then, two weeks later, plant another section. Rinse and repeat! It's like having your own personal produce subscription service, except you're both the farmer and the customer. Plus, if one batch encounters a problem (hello, surprise heatwave), you've got backup episodes—I mean, plants—already in production.

Crop Rotation: Musical Chairs for Plants

Remember that game of musical chairs from childhood? Crop rotation is similar, except instead of kids scrambling for seats when the music stops, you're moving plant families around your greenhouse to prevent soil exhaustion and pest buildup. Think of it as giving each bed a vacation from growing the same type of crop—because even soil needs a break from growing tomatoes 24/7.

The basic rule? Don't plant members of the same family in the same spot for at least three seasons.

Here's one example of a rotation: heavy feeders (e.g. tomatoes) are followed by light feeders (herbs), then soil builders (legumes). Like a very slow dance where everybody eventually gets to shake their roots in a different spot.

Soil-Free and Fancy Free: A Peek at High-Tech Growing

For the truly adventurous (you know who you are), there's always the sci-fi corner of greenhouse gardening. If you're itching to explore the cutting edge, consider hydroponics, aeroponics, and aquaponics.

Want to grow lettuce without soil? Hydroponics has you covered, letting plants party in nutrient-rich water instead. Aeroponics goes even further—imagine roots dancing in mid-air while they get misted with nutrients. And aquaponics? It's like running a fish hotel where the guests pay their rent by fertilizing your plants.

While these methods might sound like something from your favorite space colony novel, they're very real options for the curious gardener with a taste for innovation. Worth a Google rabbit hole if you're feeling experimental!

Tech Meets Dirt: Apps and Wall Calendars

Welcome to modern gardening, where your phone can tell you it's time to plant peas! Garden planning apps are like having a master gardener in your pocket, minus the dirt under their fingernails (although I've gotten my phone plenty dirty in the garden!). They can track planting dates, remind you when to succession plant, and even tell you when to start your fall crops while you're still sweating through summer.

Not into techie solutions? No judgment! A good old-fashioned wall calendar works just as well (and never needs charging). The key is finding a system that you'll actually use. Whether you're team digital or team paper, the best planning tool is the one that keeps you from finding surprise-sprouted potatoes because you forgot when you planted them.

Check-In

What a beautiful sight, friend. Your greenhouse is no longer just an empty crystal palace—it's bursting with green life.

By working through this chapter, you've gained some essential gardening knowledge. Among the many other things you know, you:

- Understand which plants make excellent "training wheels" for beginners.
- Can interpret seed packets to get vital plant info.
- Understand growing zones and how your greenhouse modifies your local climate.
- Understand basic planting techniques.
- Know some cool tricks like companion planting, crop rotation, and succession planting, for maximizing space, plant health, growing potential, and continuous harvests.
- Understand the differences between starting from seeds versus seedlings.

Let's check where your planting journey stands:

- Have you selected a few beginner-friendly plants that match your greenhouse conditions?
- Do you understand your growing zone and how your greenhouse extends your growing season?
- Have you considered how to arrange plants based on their sunlight needs?
- Do you have a basic planting calendar that accounts for seasonal changes?
- Have you decided whether to start with seeds, seedlings, or a combination of both?

Still unsure about any of these points or decisions? Not to worry. Celebrate your current level of confidence and competence. Then, make note and circle back later to the fuzzy parts. This is a journey you can take at your own pace.

Before moving on, try taking a practical step:

- Create a simple planting diagram of a few crops, showing where each crop will go based on height, light and water needs, and companion relationships.
- Pick up a seed packet, or look on the label of a seedling, and get familiar with the basic info you can find out about the plant: water and sunlight needs, spacing, planting depth, maturity and germination time, pest resistance, etc.

Quick success tip: Plant a mix of quick-yielding crops (like radishes or lettuce) alongside longer-term plants. Those fast growers will give you early wins and keep your motivation high while you wait for your slow-but-rewarding crops to mature. Nothing builds gardening confidence like harvesting something—anything!—within those first few weeks.

Wrap-Up

When you plant seeds, you're growing more than food. You're growing confidence, knowledge, and maybe even a new obsession. In the greenhouse, every day brings new growth, both for your plants and for you as a gardener.

So, go forth and plant something! Your future self will thank you with fresh, homegrown produce and a serious sense of accomplishment.

Now, as your green babies mature to reach their full-grown potential, it's time to think about the grand finale: harvesting your well-earned rewards and making sure they stick around for more than a fleeting moment.

Maybe that's the moment you've *actually* been waiting for.

In the next chapter, we'll dive into the secrets of knowing when to pluck your produce at peak perfection and how to store your garden goodies so you can savor the flavors of your success long after the last leaf has fallen. Let's go pick some fresh veggies.

9

HARVESTING AND STORAGE

MAKING THE MOST OF YOUR BOUNTY

It's an inspiring moment when all that hard work pays off. You've poured your heart and soul into your greenhouse, and now your plants are really putting on a show—strutting their stems, flaunting their foliage, and serving up nature's finest home cooking.

Really sit with that for a minute. You did it! *YOU*. You're the one who cared for these plants. Your greenhouse is transforming from science experiment to personal farmer's market—and guess what? You're both the farmer AND the lucky customer.

Harvesting time is a bit like graduation day for your greenhouse. Your plants have done their growing, and now it's time to collect your edible diplomas!

Okay, enough with the corny metaphors. But take a moment and celebrate that, out loud! (For real!) A nice "yesssss!" or "yeehaw" will do. Do a little dance. Does a body good.

Now that you're all pumped up, I'm sure you're ready to cash in on all that plant parenting.

But before you go full farmer and start taking names and plucking everything in sight, let's talk about how to harvest like a pro. Successful harvesting comes down to three important aspects: timing (catching each crop at its peak), technique, and proper tools (because nobody wants to perform surgery with a butter knife). By the end of this chapter, you'll know how to pluck produce with the best of 'em.

So, let's dive into the art of garden gathering, so you can fully enjoy the fruits of your labor.

KNOWING WHEN TO HARVEST: HOW TO TELL IT'S GO TIME

Remember when we talked about patience being a gardener's best friend? Well, here's where that virtue really pays off. Harvesting at the right moment is like catching a perfectly toasted marshmallow—wait too long, and things get messy; jump the gun, and you're left with something that's not quite ready for prime time.

When it comes to plants, harvesting too early leaves you with bland, underdeveloped flavors. Too late, and suddenly your cucumber thinks it's auditioning for the role of baseball bat. The sweet spot lies somewhere in between, and I'm here to help you find it.

Harvesting at peak ripeness isn't just about catching them before they drop; it's about catching them *at their best*. We want your tomatoes to be the *juiciest* they can be, and the *most flavorful*.

Plus, a perfectly ripe tomato isn't just juicier and more flavorful—it's actually packed with peak nutrients, like carotenoids and lycopene. The key is to catch them in their prime, just as their sugars and aromatic compounds reach perfection.

Each plant has its own way of saying, "I'm ready!" Some are obvious show-offs, like tomatoes. They're like little traffic lights, going from "go" mode to "slow down" to "stop, time to pick me!" Others play it a little more subtle.

Let's break down some common "pick me" signals:

- **Visual cues:**
    - Even coloring (like tomatoes turning fully red)
    - Size reaching typical maturity
    - Slight softening in fruits (think peaches and pears)
    - For root vegetables: Leaves starting to yellow or die back
    - For leafy greens: Harvest when the leaves are vibrant and full-sized but before they start to yellow or bolt (i.e. send up flower stalks).

- **Touch tests:**
  - Gentle squeeze test for squash and melons
  - Light tug on root vegetables (if they resist, they're probably not ready)
  - Firmness check for fruits (slight softness: not too firm, not too soft)
  - Natural detachment (e.g., melons pulling easily from the vine)

- **Smell signals:**
  - Sweet, fruity aromas from ripe melons
  - Earthy scents from mature root vegetables
  - Fragrant herbs at their peak

- **Other tests:**
  - A taste test for greens or herbs can also confirm readiness—flavor is often at its peak just before flowering.

Pro tip: If you're new to this and feeling uncertain, it's better to harvest a little early than too late. Most fruits will continue ripening off the vine, but once they're overripe, there's no turning back that clock!

Specific Signals

Here are some examples of signals from *specific* plants to give you an idea of what to look for:

- **Color changes:**
  - Tomatoes: Even coloring with no green shoulders (unless it's a green variety)
  - Peppers: Deep, rich color development (they're like mood rings for your garden)
  - Eggplants: Glossy skin with consistent color
  - Summer squash: Light, even coloring (dark spots mean it's getting bitter)
  - Beans: Bright, crisp color with no brownish spots
  - Leafy greens: Deep, vibrant color with no yellowing

- **Size and shape:**
    - Cucumbers: Full size but still slender (thick ones are usually bitter)
    - Zucchini: 6–8 inches is perfect
    - Beans: Thick as a pencil before seeds bulge
    - Peas: Pods should be plump but not bulging
    - Root vegetables: Size depends on variety (check your seed packet)
    - Broccoli: Tight, compact heads before flowers open

- **Texture tests:**
    - Melons: Slight give at the blossom end
    - Squash: Should resist thumbnail pressure
    - Root vegetables: Shoulders peeking above soil line
    - Tomatoes: Firm but with slight give (like a ripe avocado)
    - Peppers: Walls should feel firm and crisp
    - Eggplants: Skin should bounce back when gently pressed

Signs You've Waited Too Long

Because sometimes life gets busy and suddenly your vegetables are sending you desperate "Where were you?" signals:

- Zucchini: Turned into a medieval weapon (a big one)
- Cucumbers: Yellow spots appearing
- Peas: Pods looking lumpy like they're smuggling marbles
- Beans: Seeds visible through the pod
- Lettuce: Starting to stretch upward (bolting)
- Radishes: Splitting or becoming pithy
- Broccoli: Flower buds starting to open

Morning Glory: The Best Time to Harvest

Here's a secret the commercial farmers know: harvesting in the early morning gives you the best flavor and longest storage life. Why? Your plants are like teenagers—they're at their freshest after a good night's rest, fully hydrated and crisp before the day's heat sets in.

But if you're not a morning person (I feel you), here's your next best options:

1. Evening, once the day has cooled
2. Cloudy days anytime
3. Whenever you can squeeze it in (because let's be real, some harvesting is better than none)

Weather Watch: Nature's Impact

Weather can be your friend or foe during harvest season. Knowing how to roll with it is half the battle.

For instance, in colder months, plants in your greenhouse may grow slower due to reduced light, which can extend the time to maturity. Conversely, summer crops might ripen faster due to warmth and extended daylight.

Weather can further affect your timing if your greenhouse isn't sufficiently equipped for climate control (e.g. temperature). Here's how to play it smart:

- **Rain forecast:**
    - Harvest tomatoes and berries before rain to prevent splitting.
    - Let root vegetables wait (unless flooding is expected).
    - Pick leafy greens after rain (they'll be cleaner).

- **Frost warning:**
    - Harvest tender crops completely (tomatoes, peppers, eggplants).
    - Leave cold-hardy vegetables (kale, Brussels sprouts, carrots).
    - Consider emergency row covers for what you can't harvest.

- **Heat wave:**
    - Harvest early morning or late evening.
    - Pick leafy greens before they bolt.
    - Check produce more frequently (they ripen faster in heat).

Planning Ahead

Creating a harvest calendar is also a huge time-saver and headache-sparer in managing this process. By aligning your harvest schedule with plant maturity and local climate conditions, you ensure nothing goes to waste. Planning for staggered harvests can help you manage the workload, allowing you to enjoy a steady supply of fresh produce without feeling overwhelmed.

It's like making dinner reservations for your vegetables—everyone gets their turn at the table.

Know Your Plants

Want to know when your plants are ready for their big debut? The big key to harvest timing is *getting to know your plants*.

Peek at the seed packet or plant tag—it usually gives you a ballpark figure for how long the plant needs to reach maturity, often listed in days. For crops like radishes, maturity can be as fast as 20–30 days, while tomatoes or peppers may take longer.

The more you get to know your plants, the more you learn their unique signals and timing.

Keep a harvest diary (trust me). Jot down each crop's "I'm ready!" signals, what the weather was doing, and your lightbulb moments—both the brilliant and the "well, that was interesting" kind. Think of it as your personal gardening time capsule, packed with wisdom for next season's adventures. Because let's face it, memory is great but notes are better!

TOOLS OF THE TRADE: YOUR HARVESTING KIT

Harvesting tools are your plant pruning toolkit. Just like you wouldn't use a sledgehammer to hang a picture, you need the right tools for gentle harvesting.

In Chapter 3, you learned about the essential harvesting tools from our Master Tool Guide. We'll go a little more in-depth here to make your harvesting life easier. Here are some produce-picking weapons of choice:

For clean cuts (because clean cuts = happy plants):

- Sharp scissors: perfect for snipping herbs and leafy greens, allowing you to collect them without causing unnecessary damage
- Pruning shears: for thicker stems or woody branches
- Garden knife: for root vegetables (and tough stems)—its sharp edge allows for easy cutting through tough root systems without disturbing the surrounding soil too much.

Also important:

- Harvest baskets or trugs (preferably with a flat bottom), and collection containers, (nothing breaks your heart quite like dropping your fresh harvest)
- Clean gloves to protect both you and your produce

Nice-to-have additions:

- Garden apron with pockets (for tools and small harvests)
- Wheelbarrow (your back's best friend)
- Clean towels (for wiping produce)
- Labels and markers (trust me, you'll forget what you picked when)
- Long-handled fruit picker (for those hard-to-reach spots)

If you remember, we also talked tool TLC back in Chapter 3. After a few of my favorite tools got rusty, I've learned to treat my garden tools like I treat my power tools—clean 'em, oil 'em, and store 'em right. Trust me, a well-maintained tool collection will save your back and your budget.

So keep these tools clean and sharp—think of them as surgical instruments for your garden. A dull blade is like trying to cut a tomato with a spoon—messy and potentially damaging to your plants.

Tool Care (Because a Rusty Tool is a Sad Tool)

Here's a little review on how to take care of your stuff so that your stuff can take care of you.

- Clean after each use (a dirty tool spreads disease)
- Disinfect between different crops
- Sharpen regularly (dull tools damage plants)
- Store in a dry place (rust is not your friend)
- Oil moving parts monthly

THE GENTLE ART OF GATHERING

When it comes to actually picking your produce, think of yourself as a calm, skilled surgeon. Gentle movements, mindful handling, steady hands, and respect for the plant will reward you with continued harvests. Here's how:

1. Choose your moment (early morning is ideal, when plants are crisp and hydrated).
2. Twist or cut at the base of the stem rather than pulling.
3. Make clean cuts at the right spots (usually just above a leaf node).
4. Handle produce like eggs—they can bruise.
5. Have your storage containers ready nearby.

Remember: Your greenhouse isn't a one-and-done deal. Many plants will keep producing if you harvest correctly. Think of it like getting a haircut—trim properly, and it grows back healthier; hack away carelessly, and well … let's just say we've all had that one bad haircut experience.

Picking Picky Plants

Different plants need different picking approaches. Here's your cheat sheet:

- **Leafy greens:**
    - Cut outer leaves at the base.
    - Leave the growing center intact.
    - Harvest in the morning for maximum crispness.
    - Never pull—always cut or snap.
    - Take only what you need (they'll keep growing).

- **Fruiting vegetables:**
    - Use scissors/pruners for clean cuts.
    - Hold the main stem while harvesting.
    - Twist and pull gently (if not using tools).
    - Pick regularly to encourage more production.
    - Handle with care to prevent bruising.

- **Root vegetables:**
    - Loosen soil first (your back will thank you).
    - Grasp firmly at the base of the greens.
    - Pull straight up with gentle pressure.
    - Brush off excess soil (but don't wash until ready to use).
    - Check one first to gauge size.

- **Herbs:**
    - Cut stems cleanly above a leaf node.
    - Take no more than 1/3 of the plant at once.
    - Harvest before flowering for best flavor.
    - Morning harvest = maximum essential oils.
    - Use sharp scissors to prevent crushing stems.

STORAGE SOLUTIONS: KEEPING THE GOOD TIMES ROLLING

You've done it! Your harvest basket is full, and you're feeling like the ruler of your own little agricultural kingdom.

But—now what? Let's talk about keeping all this goodness *fresh*, longer than that jar of mystery condiment in the back of your fridge.

Different plants have different storage preferences (they're kind of like roommates that way). Here's a quick guide:

The Room Temperature Gang *(60–70°F/15–21°C, 50–70% humidity)*

- Tomatoes (4–7 days)
- Peppers (1–2 weeks)
- Eggplants (5–7 days)
- Winter squash (2–6 months after curing)
- Onions (2–6 months in cool, dry place)
- Garlic (3–6 months in cool, dry place)
- Citrus Fruits

Storage tip: Don't refrigerate these; they'll lose flavor faster.

The Cool Kids *(32–40°F/0–4°C, 90–95% humidity)*

- Most leafy greens (2–5 days)
- Broccoli (3–5 days)
- Brussels sprouts (3–5 days)
- Root vegetables (2–4 weeks)
- Celery (1–2 weeks)
- *Apples and pears also like it cool, but less humid (60–70%).*

Storage tip: Wrap these in slightly damp paper towels and store in perforated plastic bags in the fridge.

The Humidity Lovers

- Leafy greens (store with a damp paper towel)
- Herbs (wrapping them in damp paper towels and placing them in a breathable bag can work wonders.)
- Root vegetables (in slightly damp sand)

Storage Tips and Tricks

So you see, extending the shelf life of your produce goes beyond just tossing them in a drawer. Proper storage conditions are key to keeping your fruits and veggies in tip-top shape. You want them to stay fresh and lively, not wilted and tired.

Here are some more tips to keep them at their best:

For fruits, breathable bags or perforated containers help maintain the right balance of air and humidity, preserving their texture. You'll find they stay fresher longer, allowing you to enjoy their natural flavors without the rush of a ticking clock.

Fruits like apples and bananas produce ethylene, a gas that speeds up ripening. To avoid turning your storage area into a fast-forward aging chamber, keep ethylene-sensitive produce, such as leafy greens and tomatoes, separate. It's a bit like seating your chatty aunt away from the shy cousin at the family reunion—everyone stays in their best form.

Organizing your storage space is much like arranging a pantry; everything should have its place to maximize space and prolong freshness. Stackable bins are your allies in keeping root vegetables neat and accessible. These bins allow for airflow and prevent bruising from overcrowding.

Natural preservatives can be your secret weapon against spoilage. Here are a couple of examples:

- **Vinegar rinse for berries**:
    - A simple rinse with vinegar acts as a gentle disinfectant.
    - This helps fend off mold and extends the life of your berries.

- Think of it as a protective shield—keeping the flavor intact, no chemical aftertaste required.

- **Lime water for cut fruits**:
  - Soaking cut fruits in lime water slows down oxidation.
  - This prevents that unsightly browning and preserves their fresh flavor.
  - Your fruits stay vibrant and ready to impress.

With these tricks, your produce can outlast its store-bought counterparts, staying fresh and delicious.

And if you *really* want to preserve your harvest, there are several crafts dedicated to just that. You may be new to them, but you were once new to greenhouses, and now look at you!

PRESERVING YOUR HARVEST

Sometimes, your garden decides to give you everything at once—like when all your tomatoes ripen the day before your vacation. Don't panic! You've got options:

Freezing

Best for locking in nutrients and flavor with minimal effort:

- **Berries**: Wash, dry, and freeze in a single layer before transferring to a container.
- **Blanched vegetables**: Quick boil, ice bath, and freeze—perfect for green beans, carrots, and broccoli.
- **Herbs**: Chop and freeze in oil or water using ice cube trays for easy portions.

Drying

A classic method for compact, long-lasting storage:

- **Herbs**: Tie into bundles and hang upside down in a warm, dry place to air dry. Before you know it, you'll have fragrant, dried herbs that can spice up your dishes throughout the year. Bundle rosemary, thyme, or oregano and let time create fragrant glory.

- **Cherry tomatoes**: Halve, sprinkle with salt, and dry in the sun or a dehydrator.
- **Chilies**: String them up or use a dehydrator to make your own spice blends.

Simple Herb Drying Guide

1. Harvest herbs in morning after dew dries.
2. Remove damaged leaves.
3. Bundle 5–10 stems together.
4. Hang upside down in warm, dry place.
5. Check after 1–2 weeks.
6. Store in airtight containers.

Pro tip: Small-leaved herbs like thyme can be dried on screens or paper towels instead of hanging.

Fermenting

For tangy, nutrient-packed creations:

- **Sauerkraut**: All you need is cabbage and salt. Time and microbes do the rest, creating flavor-packed, tangy, probiotic goodness to add to any meal.
- **Pickles**: Cucumbers, vinegar, garlic, dill, and a little patience yield probiotic-rich crunchy treats.

Quick Pickling

A speedy option for smaller batches:

- **Perfect for:** cucumbers, green beans, or peppers. A vinegar brine with your favorite spices creates ready-to-eat delights in hours.

Other Methods

There are other methods like oil preservation (herbs, garlic, dried tomatoes), root cellaring (storing root vegetables, tubers, and some fruits in a cool, dark, and humid environment), and the yummiest of them all: jamming and jellying!

With jellies, you can turn your fruit harvest into delightful spreads that capture the taste of summer! Jellies are simple to make, versatile, and an excellent way to use up extra fruit.

Whatever method appeals to you, the key is to process your harvest while it's still at its peak—like capturing a perfect moment in a photograph, except this one you can eat later.

Let's not forget, of course, the cornerstone of food preservation:

Canning

Canning is an awesome skill to learn, ideal for longer storage. It might sound a bit daunting at first, but by now, you've become a master at conquering daunting tasks. With a step-by-step approach, it becomes straightforward.

There's:

- **Water bath canning**: Great for high-acid foods like tomatoes, fruits, and pickles. This method involves submerging jars in boiling water, creating a vacuum seal to keep out bacteria and spoilage.
- **Pressure canning**: Perfect for low-acid foods like beans or soups, requiring higher temperatures for safe preservation.

Here's a step-by-step preview of water bath canning:

1. **Sterilize**: Wash jars and lids, then heat to kill bacteria—think of this as giving your jars a spa day.
2. **Fill**: Pack jars with prepared produce, leaving headspace at the top for expansion. To remove air bubbles, use a non-metallic tool.
3. **Seal**: Tighten the lids, process the jars in boiling water, and let them cool. Once cooled, check the seals—the lids should not pop when pressed.

That's the *basic* process. If canning interests you, a few instructional blogs and videos, plus some supplies, will take you where you need to go.

TROUBLESHOOTING GUIDE: WHEN THINGS GO SIDEWAYS

Because, you know ... sometimes things don't go exactly as planned ...

Problem: Finding fuzzy, blue-green mold on stored produce.

Solution: Check produce daily for damage before storing, improve air circulation (don't pack them too tightly!), and adjust humidity. One moldy tomato can infect its neighbors faster than gossip at a family reunion.

Problem: Fresh-picked vegetables wilting within hours.

Solution: Start harvesting earlier in the day before the sun gets strong, improve your storage conditions (check the temperature guidelines above), and ensure that produce is cooled quickly after picking.

Problem: Fruit ripening unevenly, like tomatoes with green or yellow patches while the rest is red.

Solution: Adjust your picking schedule (they might need more time on the vine), check that plants are getting even sunlight and water, check plant health, and make sure you're not over-fertilizing.

Problem: Storage containers getting that weird slimy film on the bottom.

Solution: Clean containers with hot, soapy water between uses, add paper towels to absorb excess moisture, and make sure your produce is dry before storing (but don't wash until you're ready to use).

Problem: Discovering half your squash has turned to mush in storage.

Solution: Sort and check produce regularly (the "one bad apple" saying exists for a reason), store only perfect specimens (save the slightly damaged ones for immediate use), and maintain proper temperature and airflow.

WHY PRESERVE?

Preserving is more than practicality—it's a way to connect with the rhythm of the seasons and reduce waste while celebrating your hard work. Here's why it's worth the effort:

- **Nutritional benefits**: Retain vitamins and minerals that degrade in fresh produce over time.
- **Reduce food waste**: No more guilt about zucchini turning to mush in the fridge.
- **Save money**: A stocked pantry of homemade goods beats store-bought every time.
- **Flavor enhancement**: Preservation techniques like fermenting and pickling transform simple produce into gourmet delights.

As you line your shelves with these edible treasures, you're preserving more than food. You're bottling moments of gardening triumph and love. Each jar tells the story of a successful season. Each bite will remind you of warm days in the greenhouse and the satisfaction of growing your own food.

So grab your jars, a little salt, or some freezer bags, and get preserving. Someday, you'll be really glad you did!

The Gift of Giving Greens (and Reds and Yellows…)

And hey—if your pantry's packed, you've got some beautiful options before you start building more storage and stockpiling like you're prepping for the next Great Tomato Famine.

If there's a sparkle of generosity (or entrepreneurial spirit) in your eye, don't underestimate the joy of sharing the bounty.

Excess produce makes a perfect excuse to knock on a neighbor's door with a bag of homegrown tomatoes. They'll never forget it! Or trade zucchinis for eggs with a fellow gardener. Heck, you can even set up a charming sidewalk stand with a "pay what you can" sign and a smile.

Whether you're bartering with buddies or brightening someone's day with a surprise veggie drop-off, sharing might just be the most satisfying preservation method of all.

HARVESTING AND STORAGE

CHECK-IN

What a moment this is—your hands stained with soil, basket heavy with vibrant produce that YOU grew! There's something profoundly satisfying about that first harvest, isn't there? It's the greenhouse equivalent of watching your toddler take their first steps.

By working through this chapter, you've mastered the critical skills that transform your growing efforts into actual food on your table. You are able to:

- Recognize the visual cues, touch tests, and smell signals that tell you exactly when each crop is at its peak.
- Know which tools to use for different crops to ensure clean cuts and minimal plant damage.
- Can properly harvest different plant types (leafy greens, fruiting vegetables, root crops, and herbs).
- Understand optimal storage conditions for different produce categories.
- Have learned basic preservation methods to extend your harvest's usefulness.

Let's check where your harvesting journey stands:

- Have you assembled your essential harvesting kit with the right tools for clean, efficient gathering?
- Do you know the specific ripeness indicators for your current crops?
- Have you identified appropriate storage solutions for your different types of produce?
- Are you prepared to handle a sudden abundance with preservation techniques?
- Have you started a harvest journal to track timing and observations?

As always, there's plenty of grace as you grow. Harvesting is a skill that gets better with time and practice, just like anything else. Note which aspects you're most confident about, and which areas could use more attention.

Two things you can do before moving on, to ensure you're ready when your plants are ripe:

- Create a simple harvest schedule based on when you planted and the days-to-maturity for your crops, and
- Make sure your harvest tools and storage areas are ready and organized.

Quick success tip: Harvest in stages! Pick some crops slightly early, some at peak ripeness, and leave a few to mature longer. This "succession harvesting" not only spreads out your workload but also gives you a fascinating taste test to discover which stage you prefer. Plus, you'll learn exactly when each variety reaches its flavor peak in your specific greenhouse conditions—knowledge that no seed packet can provide!

Wrap-Up

You know what's better than growing your own food? Actually getting to eat it! And now you know exactly how to get those goodies from greenhouse to table (or freezer, or jar, or root cellar …).

Think about it: You've created this amazing space where plants thrive year-round. You've mastered the art of knowing exactly when that tomato is ripe (and resisted picking it too early—I'm proud of you!). And if you haven't yet, you will.

As you master the art of harvesting and storage, you'll find yourself part of a larger tradition—one that connects you to generations of gardeners who've celebrated the rhythm of planting, growing, and gathering. I don't know about you, but that sense of connection to the many generations before is one of the things I love most about this working-with-the-land stuff.

Breathe all that in, and let a big smile spread across your face. And take a moment to appreciate how far you've come.

You've chosen, built, equipped, and filled a greenhouse. You've planted plants in good soil, and harvested and preserved them. Remember that first day, standing in your empty greenhouse, wondering if anything would actually grow? Look at you now, probably drowning in zucchini!

Life is looking pretty good. You've come a long way.

Now, as much as I'd love to tell you, "That's all there is to it, champ, kick your feet up and relax!" I'm afraid our journey is not over yet. I should probably mention there's a slight plot twist ahead.

HARVESTING AND STORAGE

This is probably the moment you *haven't* been waiting for.

See, we're not the only ones who think your vegetables are delicious. There's a whole cast of characters—from microscopic troublemakers to surprisingly acrobatic aphids—who'd love to sample your produce before you get the chance.

That's right, I'm talking about *pests and diseases.*

Don't panic! This isn't the part where your greenhouse turns into a horror movie. Think of it more like a mystery novel where you get to be the sharp detective.

In our next chapter, we'll tackle pest and disease management with the same can-do spirit that got you this far. You'll learn how to spot the usual suspects, prevent uninvited dinner guests, and keep your harvest safe from those tiny troublemakers.

After all, you didn't grow all this goodness just to share it with some free-loading fungi or munching marauders, right? Let's make sure your hard-earned harvest stays exactly where it belongs—in your kitchen, not in some caterpillar's belly!

Remember that whole ancestral vibe we mentioned a moment ago? Soak that in again. All the victorious gardeners who've come before you have stood where you're standing and triumphed. You can as well!

And, now that you're a harvesting master, you've got even more incentive to guard those perfect peppers and delicate tomatoes you worked so hard to grow.

Ready to become a garden guardian? Let's go show those pests who's boss!

10

PEST AND DISEASE MANAGEMENT
PROTECTING YOUR GREEN PARADISE

So everything's going great. You built the greenhouse, you planted the plants. Maybe you've watched them grow, and you're even getting ready to harvest.

Well, one fine morning, you're strolling into your greenhouse, coffee in hand, ready to greet your thriving plant paradise. But wait—something's not right. Those tomato leaves are looking a bit ... nibbled. And is that *powder* on your cucumbers?

Don't panic! Every greenhouse gardener faces these challenges. Think of pests and diseases as uninvited guests at your garden party. Sure, they're annoying, but with the right strategies, you can show them the door while keeping your cool (and your plants).

CREATING A PEST SCOUTING ROUTINE: YOUR FIRST LINE OF DEFENSE

You know that moment when you walk into your greenhouse and something just feels ... off? Maybe it's a few suspicious holes in your lettuce leaves, or your tomato plants looking a bit under the weather.

Instead of waiting for these "uh oh" moments, let's set up a proper pest-scouting system that catches problems while they're still tiny enough to handle with a firm "not in my greenhouse" attitude.

The Game Plan

Pest scouting is like running a TSA checkpoint for your greenhouse—except instead of confiscating water bottles, you're looking for uninvited six-legged guests. Here's how to make it effective and, dare I say, almost fun:

1. **Map your territory**
   - Divide your greenhouse into manageable sections (I use a simple grid system).
   - Label each section clearly on a basic diagram.
   - Start at the entrance and work systematically through each zone.
   - Record-keeping tip: Note the date and findings for each section.
2. **Weekly inspection schedule**
   - Choose a consistent day and time (like Sunday mornings with coffee).
   - Allow about 15–20 minutes for a thorough inspection.
   - Check both upper and lower leaf surfaces.
   - Pay special attention to new growth and plant joints.
   - Pro tip: Set a phone reminder—because even the best intentions can get buried under life's to-do list.
3. **Essential scouting tools**
   - Good magnifying glass (10x magnification minimum).
   - Small notebook or tablet for records.
   - Clean white paper (for the tap test—more on that in a second).
   - Smartphone for photos of suspicious findings.
   - Pro tip: Keep these in a dedicated "scouting kit" so you're not hunting for tools when it's time to inspect.

Here's another tip: Teamwork makes it fun! Involve family members or helpers, assigning specific sections to each person (e.g., north quadrant for one, south for another). You can even add a competitive twist: who can spot the most aphids in ten minutes?

The Inspection Method

Here's my tried-and-true systematic approach:

1. **The overview**
   - Start with a general walkthrough.
   - Note any plants that look stressed or different from last week.
   - Check for obvious signs like webbing, droppings, or leaf damage.
2. **The close-up**
   - Examine plants in each section, starting from the bottom.
   - Check both sides of leaves, especially where leaves meet stems.
   - Look for eggs, cast skins, and actual pests.
   - Don't forget to check soil surface and pot rims.
3. **The tap test**
   - Hold white paper under branches.
   - Give them a sharp tap.
   - Watch what falls—this often reveals pests you didn't spot visually.
   - Pro tip: Do this first thing in the morning when insects are less active and more likely to drop.

Record Keeping That Actually Helps

Keep it simple but specific:

- Date and time
- Weather conditions
- Any signs of pest activity
- Location of findings
- Photos of suspicious damage
- Actions taken

Pro tip: Use your phone's voice recorder for quick notes during inspection, then transfer important details to your log later. You can use voice-to-text to keep note-taking simple.

PEST AND DISEASE MANAGEMENT

Making It Sustainable

The best pest-scouting routine is one you'll actually maintain. Here's how to make it stick:

- Start small: 15 minutes weekly is better than an abandoned elaborate system
- Combine it with other tasks: I scout while watering or harvesting
- Make it social: Turn it into a family activity or share findings with other gardeners
- Create a simple checklist to follow until it becomes habit

Remember: You're not aiming for perfection—you're creating an early warning system. Think of it like a neighborhood watch for your plants. The goal isn't to eliminate every single pest (that's neither possible nor desirable), but to catch population explosions before they happen.

When you spot something concerning, don't panic. Make a note, take a photo, and consult your pest management resources (which we'll cover in the next section). Often, early detection means you can handle problems with the gentlest possible intervention.

Also remember: A healthy plant is your best defense against pests and diseases. Keep your plants strong through proper care, and they'll be better equipped to fight off invaders on their own.

MEET THE USUAL SUSPECTS: COMMON GREENHOUSE PESTS

Let's get to know our unwanted visitors. These tiny troublemakers see your greenhouse as an all-you-can-eat buffet, complete with climate control and cozy corners to hide in.

The Most Wanted List

Aphids: These tiny sap-suckers are the rabbits of the insect world—they multiply faster than social media trends. They're small, soft-bodied, and come in various colors (though I doubt they're trying to be fashionable). They often leave behind a sticky substance called honeydew, which can attract sooty mold. Look for:

- Curled, wilting leaves
- Sticky residue on leaves
- Clusters of tiny insects, especially under leaves
- Distorted plant growth

Spider mites: These are microscopic spiders that have web-spinning parties on your plants. They thrive in warm, dry conditions and can turn your green haven into their personal yoga studio. Watch for:

- Fine webbing between leaves
- Tiny specks moving on leaves
- Yellow or bronze leaf discoloration
- Stippled appearance on leaves

Whiteflies: These tiny white-winged insects are like annoying party crashers who bring all their friends. They feed on plant sap, and like aphids, they also leave behind sticky honeydew that can lead to sooty mold—think plant blackheads, but worse. Signs include:

- Clouds of tiny white insects when plants are disturbed
- Sticky residue on leaves
- Black, sooty mold growth
- Yellowing leaves

Thrips: These slender, tiny insects puncture plants to feed. These speedsters are like tiny race cars zooming around your plants, leaving silvery damage in their wake. Look out for:

- Silvery streaks on leaves
- Distorted growth
- Black specks (their droppings)
- Scarred or deformed flowers

Here are a few others:

Fungus gnats: Tiny flies that lay their eggs in damp soil. The larvae nibble on plant roots, which can lead to plant health issues, like stunted growth.

Caterpillars and leaf miners: These pests chew through leaves or burrow within them, leaving behind visible trails or holes.

PEST AND DISEASE MANAGEMENT

Slugs and snails: These pests are more common in humid greenhouses. They feed on foliage, leaving irregular holes and a telltale slimy trail.

Mealybugs: Cotton-like white clusters on stems and leaves indicate these pests, which suck sap and weaken plants.

Scale insects: Tiny, shell-like insects that latch onto plant stems and leaves, causing yellowing and a general decline in plant vigor.

Becoming a Plant Detective

The key to managing these pests is early detection. Think of yourself as a horticultural Sherlock Holmes, complete with magnifying glass (yes, really—get one!). Here's your investigation toolkit:

1. **Weekly inspections:** Make a ritual. Coffee in one hand, magnifying glass in the other.
2. **Sticky traps:** Yellow sticky cards are like tiny crime scene investigators, capturing evidence of flying pests.
3. **Documentation:** Keep a pest diary. It sounds nerdy, but tracking outbreaks helps you prevent future problems.
4. **The shake test:** Hold a white paper under suspicious leaves and gently shake them. If tiny specks start moving, you've got company.

Top Signs of Pests

- Holes or ragged edges in leaves, indicating chewing insects like caterpillars or beetles
- Curling, distorted, or puckered leaves—often a sign of aphids or thrips
- Discolored spots or stippling on leaves, suggesting spider mites or scale insects
- Sticky residue (honeydew) on leaves or underneath plants, typically from sap-feeding pests
- Sooty black mold growing on honeydew deposits, a secondary sign of pest infestation
- Wilting or yellowing leaves despite adequate water, which may indicate root-feeding pests
- Fine webbing between leaves and stems—a classic spider mite calling card

Pro tip: Check the undersides of leaves regularly, as many pests prefer to hide there. Early detection means easier control.

NATURAL PEST CONTROL: YOUR GREEN DEFENSE ARSENAL

Before you reach for the hazmat suit and chemical warfare options, let's talk about gentler solutions that work *with* nature rather than against it.

Your Organic Armory

Neem oil: Nature's pest control superhero. This oil, pressed from neem tree seeds, is like kryptonite for many insects. It disrupts their feeding and breeding cycles without harming beneficial insects.

Diatomaceous earth: Imagine microscopic seashells that act like tiny razor blades to insects' exoskeletons. It's deadly to pests but harmless to humans and pets.

Insecticidal soaps: These specialized plant-based soaps work like molecular pickpockets. They target soft-bodied bugs like aphids and spider mites by dissolving their outer coating, causing them to dehydrate and die. Simply spray directly on the unwanted visitors, and the soap disrupts their cell membranes. Think of it as natural pest control that works through basic chemistry rather than toxins.

Note: While organic pesticides from the store shelf are derived from natural sources, it's important to recognize that 'natural' doesn't always mean 'safer' if you're buying something pre-made. Some organic pesticides can be quite toxic. Always *research the stuff you use,* evaluate each pesticide's properties and use them responsibly, whether they're synthetic or organic.

DIY Pest Control Solutions:

- **Garlic-chili spray**
    - Blend 4–5 garlic cloves
    - Add 1–2 hot chilies
    - Steep in 2 cups water overnight
    - Strain and spray. Think of it as hot sauce for your plants—except they wear it instead of eating it.

PEST AND DISEASE MANAGEMENT

- **Soap spray**
    - 1 tablespoon mild liquid soap
    - 1 quart water
    - Optional: Add a few drops of neem oil. It's like giving pests a slippery slide right off your plants.

Store-Bought Pest Control

Note: Here's a reality check about organic pesticides: Just because something's labeled "organic" or "natural" doesn't automatically make it gentle as a summer breeze. Think about it—poison ivy is natural too, but you wouldn't want to roll around in it!

Store-bought organic pesticides can pack quite a punch, and some might be harsher than you'd expect. So before you spray anything on your precious plants (organic or otherwise), do your homework and read those labels. And put on those gloves too! Your hands, plants, and local beneficial insects will thank you!

Speaking of beneficial insects ...

Beneficial Insects

Turn your greenhouse into nature's version of The Avengers by introducing:

Ladybugs (aphid terminators): A single adult can devour up to 50 aphids daily, making them efficient pest controllers. Release at dusk near aphid-infested plants for best retention.

Predatory mites (spider mite nemeses): These specialized hunters consume multiple spider mites daily. They are particularly effective in humid greenhouse conditions. Introduce them at the first sign of spider mite damage.

Parasitic wasps (caterpillar controllers): These microscopic wasps target pest eggs, preventing caterpillar problems before they start. Each wasp can parasitize up to 100 pest eggs.

Lacewings: These delicate-looking predators pack a punch. Their larvae (nicknamed "aphid lions") can consume up to 200 aphids or other soft-bodied pests per week. They're particularly valuable because they stay in your greenhouse longer than ladybugs.

Praying mantids: These charismatic generalist predators handle larger pests like moths and beetles. One mantid can clear a remarkable territory of harmful insects, though they don't discriminate between pest and beneficial insects.

Pro tip: Release beneficial insects when pest populations are present but still manageable. They need some prey to establish themselves. Think of it as providing them with a starter meal so they'll stick around for the main course.

Consider creating "insect hotels" with rolled cardboard or bundled hollow stems to provide shelter for your beneficial insect workforce between hunting sessions.

Plants That Keep Beneficial Insects on Duty

Think of these plants as your beneficial insects' favorite bed & breakfast spots—they provide both food and shelter to keep your garden allies hanging around.

Here are some examples of how to create an insect-friendly greenhouse:

- **Sweet alyssum:** Provides nectar for parasitic wasps and creates low-growing cover
- **Dill and fennel:** Their umbrella-shaped flowers attract ladybugs and lacewings with accessible nectar
- **Oregano:** When allowed to flower, attracts multiple beneficial species with its clustered blooms
- **Marigolds:** Beyond pest deterrence, their pollen supports predatory mites
- **Yarrow:** Its flat-topped flower clusters serve as landing pads for beneficial wasps
- **Calendula:** Produces abundant nectar and sticky stems that trap pests, creating hunting grounds for beneficial insects
- **Buckwheat:** Quick-growing cover crop that produces nectar-rich flowers in just 3–4 weeks
- **Borage:** Continuous bloomer providing reliable nectar sources throughout growing season
- **Nasturtiums:** Double duty as both trap crop and nectar source

Pro tip: Plant these in clusters rather than single specimens. Think "insect pit stops" rather than "isolated gas stations." Maintain a rotation of blooming plants to ensure continuous food sources. Consider dedicating about 10% of your greenhouse space to these beneficial-attracting plants for optimal pest management support.

PLANT DISEASES: WHEN YOUR PLANTS NEED A DOCTOR

Just like us, plants can catch their share of bugs (not the bugs we just talked about, but the pathogenic kind). Learning to spot the symptoms early is key to keeping your greenhouse healthy.

Common Diseases and Their Symptoms

- **Powdery Mildew:**
    - Looks like someone dusted your plants with flour
    - Starts as small white spots
    - Can spread rapidly in warm, dry conditions
    - Particularly loves cucumbers and squash
- **Blight:**
    - Brown or black spots on leaves
    - Can spread to stems and fruits
    - Often worse in humid conditions
    - Tomatoes and potatoes are favorite targets
- **Root Rot:**
    - Wilting despite moist soil
    - Yellowing leaves
    - Brown, mushy roots
    - Often caused by overwatering

Prevention Is Better Than Cure

1. **Air circulation:** Keep the air moving with fans. It's like your greenhouse's HVAC system.
2. **Plant spacing:** Give everyone breathing room. No plant cuddle puddles!
3. **Watering habits:** Water the soil, not the leaves.
4. **Clean tools:** Sanitize your pruners between plants. Would you use the same fork at a buffet without washing it?

Also, select disease-resistant varieties whenever possible.

Think of it as choosing a car with built-in safety features rather than trying to add them later. Many modern cultivars are bred with natural resistance to common greenhouse diseases like powdery mildew, blight, or fusarium wilt.

Check those seed catalogs and plant labels for resistance codes—they're like your plants' immune system report card. For example, when you see "VFN" after a tomato variety, that's shorthand for built-in protection against Verticillium wilt, Fusarium wilt, and Nematodes. A little research during selection can save you a lot of treatment later.

Early Detection and Response

Time to become a plant doctor. Your green buddies provide all kinds of hints when they are not at their best.

Here are the hidden messages to look out for:

- Unusual leaf spots or discoloration
- Wilting even when there is sufficient watering
- Powdery or fuzzy growth on leaves
- Stem lesions or rotting
- Yellowing leaves in unusual patterns

Here's how to respond if disease strikes:

1. Remove and destroy infected plant parts immediately.
2. Improve air circulation around affected areas.
3. Adjust watering practices to avoid overhead spraying.
4. Consider organic fungicides if appropriate.
5. Document the outbreak to prevent future issues.

Think of disease management like airport security—your first line of defense is prevention, but you need a solid response plan for when something slips through. Regular monitoring and swift action are your best allies in keeping your greenhouse healthy.

PEST AND DISEASE MANAGEMENT

PEST PREVENTION: YOUR PROACTIVE PROTECTION PLAN

If you're feeling bogged down by all this maintenance and pest and disease prevention, it might help to think of it like maintaining your car. You can perform regular maintenance and avoid problems, or you can wait until that mysterious engine noise becomes a roadside breakdown. Trust me, I've learned this difference the hard way, both in my garden and my car!

Your greenhouse operates on the same principle. Oil changes keep the engine on your car running smoothly, and checkups on your greenhouse keep it running smoothly.

And when it comes to controlling pests and saving your plants (the stars of your greenhouse show), the best offense is a good defense.

So here's how to make your greenhouse less appealing to unwanted guests:

Structural Strategies

1. **Screens and Barriers:**
    - Install fine mesh over vents.
    - Use door strips to prevent entry.
    - Seal any gaps or cracks.
    - *(Think of this as home security for your plants.)*

2. **Environmental Controls:**
    - Maintain proper temperature.
    - Control humidity levels.
    - Ensure good air circulation.
    - *(Your greenhouse should be comfortable for plants, not pests.)*

Cultural Practices

Crop rotation:

- Don't plant the same crops in the same spot the following year.
- Mix up your plant families.
- Keep records of what grew where.

Companion planting:

Many companion plants are known for their ability to repel pests due to their scent, chemical compounds, or physical characteristics.

Here are some popular examples of plants, and pests they help fight against:

- Basil (mosquitoes, flies, and aphids)
- Mint (ants, aphids, cabbage moths, and fleas)
- Rosemary (cabbage moths, carrot flies, and mosquitoes)
- Thyme (whiteflies and cabbage loopers)
- Dill (aphids, spider mites, and squash bugs)
- Chives (carrot flies and aphids)
- Marigolds (nematodes, whiteflies, aphids, and mosquitoes)
- Nasturtiums (aphids, squash bugs, and whiteflies)
- Calendula (aphids … and attracts predatory insects like ladybugs)
- Lavender (moths, fleas, and mosquitoes)
- Garlic (aphids, Japanese beetles, and spider mites)
- Onions (carrot flies and aphids)
- Radishes (cucumber beetles)
- Leeks (carrot flies)
- Borage (tomato hornworms and cabbage moths)
- Tansy (ants, flies, and beetles)
- Fennel (slugs and snails)
- Catnip (fleas, ants, aphids, and squash bugs)

A well-planned greenhouse is like a fortress with a flourishing food court—unwelcome pests get blocked at the gates, while beneficial insects enjoy the five-star amenities you've thoughtfully provided.

Things won't be perfect. You may not be able to have a "pest-free" greenhouse, but you can certainly shoot for pest-resistant.

Your goal isn't to create an impenetrable bubble, but rather a balanced ecosystem where the good guys have the upper hand. With these preventive measures and strategic plantings in place, you're well on your way to maintaining a thriving, pest-resistant greenhouse that works smarter, not harder.

PEST AND DISEASE MANAGEMENT

Check-In

Look at you, mighty plant protector. Once, you were a mere greenhouse builder. Now, you are a vigilant defender of your leafy kingdom. Those pests might see your greenhouse as an open buffet, but you're now armed with the knowledge to show them the door.

By working through this chapter, you've gained some seriously valuable skills. You:

- Understand how to create and maintain a regular pest scouting routine—your early warning system against unwanted visitors.
- Can identify common greenhouse pests like aphids, spider mites, whiteflies, and thrips by their distinctive damage patterns.
- Know natural pest control options from neem oil to beneficial insects that work with nature rather than against it.
- Recognize major plant diseases and their telltale symptoms before they spread throughout your greenhouse.
- Understand preventative strategies like proper air circulation, plant spacing, and strategic companion planting.
- Can implement structural defenses like screens and barriers to create a fortress for your plants.

Let's check on where you are in your plant protection journey:

- Have you set up a pest scouting routine?
- Do you have any basic pest control supplies ready?
- Have you identified potential companion plants that could help with common pests?
- Have you adjusted your watering practices to help prevent disease (watering soil, not leaves)?
- Have you installed any physical barriers like screens on vents to prevent pest entry?

Remember, it's normal to be unsure on some of these topics. As you read over this list, check what needs revisiting and further research, and give yourself a pat on the back for each small step you've taken. Every pest you identify is building your expertise, and every preventative measure you take will lead to a healthier greenhouse.

Before moving on, try creating a simple pest scouting routine, and think of how you might group it with other greenhouse routines (watering, pruning, etc). You can always change it later, but think about the best times that work for you. This thought process will help you in the next chapter, too!

Quick success tip: When you spot a pest issue, take a photo before treatment, then another a week later to document your progress. These before-and-after snapshots not only help you track what's working, but they're also incredibly satisfying proof of your growing pest management skills. Plus, when that gardening friend asks for advice on aphids, you'll have visual evidence that you're pretty much a pro at this.

Wrap-Up

As we wrap up our pest and disease management strategies, remember that every gardener faces these challenges. What matters is how you respond. You are a mighty plant protector, remember? You're armed with knowledge, natural solutions, and a keen eye for trouble.

And hey, if some sneaky pest manages to claim a plant or two (or ten) along your journey, consider it their way of teaching you a valuable (if slightly annoying) lesson. After all, every master gardener's story includes a few chapters about the ones that got away!

Next up, we'll explore the nuts and bolts of keeping your greenhouse running smoothly with regular maintenance and troubleshooting tips. Because just like any home, your greenhouse needs regular TLC to stay in tip-top shape!

11

MAINTENANCE AND TROUBLESHOOTING

YOUR GREENHOUSE'S HEALTH & WELLNESS PLAN

If you think maintenance sounds about as exciting as watching paint dry, I get it! But here's the thing: maintaining your greenhouse is less like boring chores and more like being a spa director for your plants. Think of it as running a wellness retreat where every plant gets the five-star treatment it deserves.

And troubleshooting? Well, that's just being a plant detective. And who doesn't love a good mystery? (Especially one where the culprit usually turns out to be something fixable like "oops, forgot to open the vent" rather than some diabolical master criminal.)

If quirky contexts don't do it for you, remember the prize ahead. Gardening can be pretty addicting, and nothing motivates quite like success. Once you taste that first sun-warmed tomato or snip those fresh herbs for dinner, you'll find yourself practically inventing excuses to pop in and say hello to your leafy friends!

MAINTENANCE TO KEEP YOUR PLANT SPA SPIC-AND-SPAN

Just like you probably have a morning routine (stumble to coffee maker, achieve consciousness, remember you're a person), your greenhouse needs its own daily rituals. Here's what keeps your plant paradise running smoothly:

Morning Check-In

Start each day with a quick wellness check. It's like being a doctor doing rounds, except your patients photosynthesize:

- Do a visual inspection. (Any plants looking dramatic?)
- Check temperature and humidity. (Remember those ideal ranges we talked about in Chapter 4.)
- Ensure ventilation is working. (Your plants need their morning fresh air too.)
- Look for any overnight shenanigans from pests

Pro tip: Make this part of your morning coffee routine. Your plants get attention, and you get to enjoy your coffee in a tropical paradise. Win-win!

Weekly Wellness Program

Think of these as your greenhouse's weekly spa treatments:

1. **Deep Clean Sunday** (or whatever day works for you)
   - Sweep paths.
   - Wipe down surfaces.
   - Clean any algae build-up (it's like that green stuff that shows up on shower tiles, but outdoors).
   - Sanitize tools.
2. **Systems Check**
   - Inspect irrigation systems for clogs or leaks.
   - Check fans and vents are operating smoothly.
   - Look over structural elements (any loose screws getting rebellious).
3. **Plant Salon Day**
   - Prune dead or yellowing leaves (like a haircut, but for plants).
   - Check for any signs of disease.
   - Rotate plants if needed (everyone deserves their time in the spotlight).

MAINTENANCE AND TROUBLESHOOTING

Monthly Maintenance Mastery

Monthly tasks are like your greenhouse's quarterly performance review, just more frequent and with less awkward small talk:

1. **Structure Survey**
   - Check all seals and weatherstripping.
   - Inspect for any developing leaks.
   - Look for loose panels or hardware.
   - Test all automatic systems.
2. **Deep Cleaning Detail**
   - Thoroughly clean all surfaces.
   - Disinfect tools and equipment.
   - Clean inside and outside of panels (your plants deserve a room with a view).
3. **Systems Overhaul**
   - Clean and maintain ventilation systems.
   - Flush irrigation systems.
   - Check and calibrate monitoring equipment.

SEASONAL SHIFTS: HELPING YOUR GREENHOUSE ADAPT

Each season brings its own challenges, like having four different roommates move in throughout the year. One's too hot, one's too cold, one keeps dropping leaves everywhere, and one can't stop crying. (No offense, Spring, you're still my favorite!) Here's how to help your greenhouse accommodate each seasonal guest:

- **Spring prep**
  - Clean and repair any winter damage.
  - Check and repair screening.
  - Prepare ventilation systems for warmer weather.
  - Clean panels thoroughly to maximize light.

- **Summer ready**
  - Install shade cloths before the real heat hits.
  - Check irrigation systems are ready for heavy use.
  - Ensure all ventilation is working perfectly.
  - Double-check structural stability (storms can pack a punch).

- **Fall transition**
    - Clean and store shade cloths.
    - Check heating systems before you need them.
    - Seal any drafts.
    - Clean gutters and drainage systems.

- **Winter prep**
    - Insulate where needed.
    - Check heating systems again.
    - Reinforce structure if needed for snow load.
    - Have emergency supplies ready (because winter storms don't check your calendar before arriving).

THE DETECTIVE'S GUIDE TO TROUBLESHOOTING

When something's off in your greenhouse, it's time to channel your inner Sherlock Holmes. Here's how to solve the most common greenhouse mysteries:

Strange Plant Behavior

If your plants are acting weird (and not in the cool way), check these usual suspects:

1. **The Wilting Wonder**
    - First check: Soil moisture (too much? too little?)
    - Second check: Temperature (is it sauna day in there?)
    - Third check: Root health (happy roots = happy plants)
2. **The Mysterious Yellowing**
    - Check nutrient levels
    - Look at watering patterns & evidence of water levels
    - Consider light exposure
    - Investigate root zone
3. **The Growth Slowdown**
    - Temperature check
    - Light levels
    - Nutrient availability
    - Season (sometimes plants just need a nap)

MAINTENANCE AND TROUBLESHOOTING

Getting Down to the Root

I've mentioned checking root health a few times, and you may wonder what that entails. Well, it's a bit like giving your plant a gentle wellness exam—you want to be thorough without making them uncomfortable. Here's how to play plant doctor without causing trauma:

For potted plants, gently tip the pot sideways (supporting the plant's stem with one hand like you're dipping your dance partner), and ease the root ball partially out. Healthy roots should look firm and light-colored—think fresh angel hair pasta, not overcooked spaghetti. If they're dark, mushy, or smell like that container of mystery leftovers in the back of your fridge, you've got root rot on your hands.

For plants in the ground, you can do a little archaeological dig near the plant's drip line (that's the outer edge where the leaves extend). Use your fingers or a small trowel to carefully expose a few roots—like uncovering dinosaur bones, but gentler. Just remember to tuck them back in when you're done snooping!

Pro tip: If you're nervous about disturbing your plants (totally normal—we all get first-time root-check jitters), start with a small, sturdy plant. Herbs like thyme or sage are usually pretty chill about the whole process. Save the root examination of your prize-winning tomatoes for when you've got more confidence in your plant-handling skills!

Structural Surprises

When your greenhouse itself is acting up:

1. **Leaks**
   - Check seals around joints.
   - Inspect panel connections.
   - Look for condensation vs. actual leaks.
   - Check guttering and drainage.
2. **Temperature maintenance checklist**
   - Weekly: Check thermostat calibration.
   - Monthly: Inspect insulation integrity.
   - Seasonal: Test climate control systems.
   - *For detailed climate control information, see Chapter 4.*

3. **Ventilation problems**
   - Clean fans and vents.
   - Check for obstructions.
   - Verify power connections.
   - Test automatic systems.

Your Troubleshooting Toolkit

Keep these essentials handy:

- Basic tool set
- Spare parts (especially common ones like clips and seals)
- Repair tape
- Cleaning supplies
- First aid kit (for you, not the plants)
- Emergency contact list
- Manual for your greenhouse
- This book! (shameless, I know)

Prevention: Your Secret Weapon

The best way to handle problems is to prevent them in the first place (revolutionary concept, I know). Here's your prevention toolkit:

1. **Keep a greenhouse journal:**
   - Track what works.
   - Note what doesn't.
   - Record maintenance dates.
   - Write down weird things you notice.
2. **Create maintenance checklists for:**
   - Daily tasks.
   - Weekly must-dos.
   - Monthly maintenance.
   - Seasonal prep.
3. **Build an early warning system:**
   - Learn how your plants say, "uh, something's not right."
   - Know your greenhouse's normal sounds and smells.
   - Stay ahead of seasonal changes.
   - Trust your gut when something seems off.

MAINTENANCE AND TROUBLESHOOTING

When to Call in the Cavalry

Sometimes, you need backup, and that's okay!

Consider professional help when:

- Structural issues appear
- Electrical systems act up
- Major leaks develop
- You're just not sure what's wrong

Remember: Even master gardeners sometimes need help. There's no shame in asking an expert!

Don't forget about gardening groups, both local and online. There are plenty of communities, forums, and blogs all over the internet, brimming with support, answers to questions, and shoulders to cry on. If you look around a little, it's not hard to build yourself a personal gardening cheerleading squad, complete with virtual high-fives and an endless supply of "been there, killed that plant too" solidarity!

CHECK-IN

While we're on the subject, take this big virtual high-five from me. Wow. Seems like just yesterday, we were sitting there on your back porch, with a book and a dream. Well, you were. I was in the book, kind of ...

Anyway, now here you are, having made the dream happen, and keeping it alive and thriving under your attentive care. By working through this chapter, you've developed a maintenance mindset that'll keep your glass oasis running strong for seasons to come. Here are just some of the things you can do now:

- Run wellness checks: Spot plant issues early by monitoring temperature, humidity, airflow, and pest shenanigans.
- Keep a weekly maintenance routine: Clean surfaces, sanitize tools, and check systems like irrigation and ventilation.
- Handle monthly and seasonal tune-ups: Inspect structural elements, clean panels, prep for seasonal shifts, and keep your greenhouse weather-ready.

- Troubleshoot common plant problems and greenhouse structural issues.
- Maintain greenhouse systems.
- Identify when to DIY and when to call in the professionals.

Let's check where your greenhouse maintenance journey stands:

- Have you created a simple maintenance schedule tailored to your greenhouse?
- Do you have a basic troubleshooting toolkit assembled and easily accessible?
- Are you keeping a greenhouse journal to track what works and what doesn't?
- Are you familiar with your greenhouse's "normal" sounds, smells, and functioning?
- Have you planned for seasonal transitions and the different challenges each brings?

Not checking all these boxes yet? No sweat! Learning your greenhouse takes time, and every maintenance task you perform build competence and confidence.

Your plants don't need perfection—just your attention and growing expertise.

Before moving on to our final chapter, try these concrete steps:

- Create a simple one-page maintenance checklist (daily/weekly/monthly) to post in your greenhouse.
- Assemble or plan a basic repair kit with tape, clips, ties, and tools specific to your greenhouse type.

Quick success tip: Label everything in your greenhouse toolkit and maintenance supplies with bright, waterproof tags. When you're dealing with a sudden leak during a downpour or need that special connector in a hurry, you'll save precious minutes not having to dig through mystery boxes. Future frazzled you will send current organized you a thank-you note!

MAINTENANCE AND TROUBLESHOOTING

Wrap-Up

Maintenance might not be the most glamorous part of greenhouse gardening, but it's what keeps your plant paradise thriving. And honestly? There's a lot of purpose you can find in it, too. There's real satisfaction in giving back to the space that fills your table—and your soul—with so much goodness. Every task, big or small, is an investment in your greenhouse's future and your growing success.

Remember as you build, fix, tweak, repeat ... the goal isn't perfection—it's progress. Every challenge you solve adds to your greenhouse wisdom.

And speaking of progress and greenhouse wisdom, let's take a look at yours. Sometimes, you've got to pause the climb and look down the mountain to see how far you've come.

Look at all the things you've overcome so far. Think back to the obstacles that overwhelmed you, but you found a way over them. You did that! And guess what? You'll encounter more, and you'll overcome them too.

Let that sink in. And then, I want you to think about something else and let it sink in, too:

Remember how we were talking about asking others for help? Well, before long, *you'll* be the one giving out the wisdom. Let me tell you, it's a pretty rewarding feeling to help someone else, especially when you know what it's like to be in their shoes.

Before you know it, you'll be able to spot that frazzled look in another beginner's eye and think, "I remember that!" That's when you'll know, "This is my moment." It'll be *your* turn to step up and be someone else's greenhouse garden guide.

Someday soon (if not already), you'll be the one passing the trowel of knowledge to the next wide-eyed, dirt-smudged dreamer who wanders into your little slice of greenhouse paradise, ready to learn from the best. (That's you, by the way. In case that wasn't clear. Wink, wink.)

So keep at it, you tenacious, green-thumbed warrior. Because you have an important part to play in the great circle of life (cue dramatic music). There's a special kind of joy that comes from sharing your hard-earned know-how and watching a new gardener's face light up like a sunrise with relief and renewed enthusiasm. That's the undeniable satisfaction of helping things grow—plants and people alike.

Well, in case I haven't made it clear yet in my inspirational speech (that is way too drawn out for the end of a maintenance chapter), let me say it again: *the growth in this journey extends far beyond the walls of your greenhouse.*

In our next (and final!) chapter, we'll explore how the lessons you've learned under glass can transform your life beyond the potting bench.

Get ready to dig deep and discover how cultivating plants can also help you cultivate a thriving community, a grounded sense of purpose, and a more joyful, connected way of living. Let's get growing, inside and out!

A QUICK NOTE FROM JR
(BEFORE WE GET ALL PHILOSOPHICAL)

You know that feeling when you've just figured out something really cool and *have* to tell someone about it? Like when you discovered you could diagnose plant problems by their leaf spots (and promptly became the neighborhood "plant detective"), or when you mastered the art of perfect ventilation timing? That rush of "I know what I'm doing!" is pretty amazing, right? And I bet you've had quite a few of those moments by now.

If you're finding value in our little gardening heart-to-hearts (and I hope you are!), would you consider spreading some of that greenhouse goodness around by leaving your review for this book online? It doesn't take much, and it has huge returns. Seriously, even a small sentence or two is powerful enough to change the WORLD! Okay, maybe that's extreme, but it does help a ton, and it only takes 1–2 minutes of your time!

Just like sharing surplus tomatoes with neighbors or passing along successful growing tips, leaving a quick review helps other aspiring greenhouse gardeners find their way to this guide. Think of it as planting a seed of knowledge that'll grow into someone else's thriving garden! You can find the review section on the website where you got this book. Again, it takes just a couple of minutes, but like that first seedling you carefully tended, it can grow into something pretty spectacular.

Scan here to review this book on Amazon!

Now, speaking of growth (smooth transition, right?), let's dig into some of that deep philosophical soil. Because while we've been busy nurturing plants, something else has been quietly flourishing too—*you!* Let's explore how your greenhouse journey is growing more than just vegetables ...

12

BLOOMING BEYOND THE GREENHOUSE

HOW PLANTS GROW PEOPLE & COMMUNITIES

There's a special feeling you get when you look at old garden photos and suddenly realize your "tiny" seedlings have turned into a jungle that would make Tarzan feel at home. It's like looking back at an old middle school yearbook and feeling that rush (nostalgia, embarrassment and everything in between). You find yourself cringing, laughing, and maybe even shedding a tear.

Growth happens so gradually that sometimes we don't notice it until we take a step back. And here's a fun secret: it's not just your plants that are growing. You're growing too! And, not in the "Oops, I ate too many homemade pizzas" way, but in the, "Hey, when did I become the person my neighbors ask for gardening advice?" way.

Building a greenhouse grows more than just plants—it cultivates some pretty impressive human traits, too! You develop patience (because you can't exactly microwave a seed into a salad), problem-solving creativity (hello, MacGyver-level vent repairs), and resilience (sometimes plants die, and that's okay! Deep breaths. Just means more compost, right?)

Your observational skills sharpen as your eyes open to those tiny leaf changes and suspicious spots. And let's not forget adaptability—because nature loves throwing curveballs, and you learn to dance with them all.

Before you know it, you've grown this beautiful blend of humility and confidence, becoming the neighborhood "plant whisperer". And you came by it through honest, down-to-earth work! You and I both know (along with your journal) of the dead plants it took to learn what *not* to do.

So it's time to reflect and get inspired. And speaking of your journal …

THE POWER OF REFLECTION: YOUR GARDEN'S STORY

Think of keeping a gardening journal like creating a time-lapse video of your journey, except instead of just watching plants pop up like magic, you're capturing the whole adventure—triumphs, whoopsies, and those "what was I thinking?" moments that turn into valuable lessons.

Why Journal?

- Track what worked (and what definitely didn't).
- Notice patterns you might miss otherwise, like how that one corner of your greenhouse stays cooler, and your lettuce likes it there. Or how your tomatoes grew especially plump when you sang to them (hey, I'm not doubting you).
- Create your own personalized growing guide.
- Celebrate progress. (Yes, those adorably wonky first tomatoes count!)
- Build confidence as you see how far you've come.

Making It Your Own

Your journal can be as unique as your garden. Maybe you're a tech enthusiast who loves garden planning apps, or perhaps you're old-school, preferring the satisfying scratch of pen on paper. Either way, the goal is to create a record that works for you. Include:

- Planting dates and harvest times
- Weather patterns and their effects
- Photos of your greenhouse's transformation
- Sketches of your layout (artistic talent optional!)
- Notes about what different plants seem to like (or protest against)
- Those "aha!" moments when something finally clicks

Journal Prompts to Get You Growing

Sometimes you want to journal, but when you sit down to do it, you find yourself staring at a blank page. Here are some prompts to get those thoughts flowing:

- "What surprised me most this season?"
- "What would I tell my beginner gardening self?"
- "My biggest challenge was ... and here's how I tackled it ..."
- "What was the most rewarding aspect about this season?"
- "What unexpected discoveries did I make about my garden or myself?"
- "Today in the greenhouse, I noticed ..."
- "Next season, I want to try ..."

YOUR GREENHOUSE, YOUR SANCTUARY

Now, let's discuss something seed packets never tell you. Your greenhouse may be place for growing plants, but it's not just that. It's also a retreat for growing peace of mind.

I mean, have you ever noticed how your shoulders drop about three inches as soon as you step inside?

There's a certain joy I can't quite put into words. Something about that humid air and earthy smell that makes the day's stress melt faster than a popsicle in July.

Creating Your Happy Place

Transform a corner of your greenhouse into your personal chill spot. Add:

- A comfy chair (garden dirt brushes off, but a sore back is no fun)
- Small table for your beverage of choice
- Maybe a little fountain for those soothing water sounds
- Wind chimes or gentle background music, if you're feeling fancy
- Fragrant plants nearby (like lavender, mint, or jasmine). You've got the setup! Might as well cultivate some natural aromatherapy.

Mindful Moments

Try these simple mindfulness practices while gardening:

- Focus on the sensation of the soil in your hands.
- Listen to the different sounds in your greenhouse.
- Notice the various shades of green around you.
- Take slow, deep breaths of that amazing greenhouse air.
- Appreciate the small miracles. (Hello, first sprout!)
- Match your breathing to your actions, creating a flow that calms the mind and invigorates the body.

BUILDING YOUR GARDEN COMMUNITY

Remember thinking you were the only one who talked to their plants? Surprise! There's a whole world of us plant people out there, and connecting with them is like finding your long-lost garden family. These connections aren't just fun—they're like adding fertilizer to your gardening knowledge!

Finding Your Plant People

- Local gardening groups (where plant nerds unite!)
- Online communities (global garden friends at your fingertips)
- Neighborhood seed swaps (trading is the new shopping)
- Community gardens (because more gardens = more joy)
- Social media plant groups (where it's totally normal to post 47 pictures of your first ripe tomato)

Sharing the Abundance

Remember that time you had so many zucchini you considered leaving anonymous vegetable donations on doorsteps? There's a better way! Create a neighborhood exchange system:

- Organize seasonal seed swaps.
- Trade surplus produce. (Cucumbers for carrots, anyone?)
- Share tools. (Not everyone needs their own rototiller.)
- Exchange gardening tips over the fence.
- Offer cuttings and divisions from your favorite plants.

Hosting a Greenhouse Open House

Ready to show off your green paradise? Hosting an open house is like throwing a party where your plants are the guests of honor! Here's how to make it awesome:

- **Pick a good time:** Choose when your greenhouse is looking its best.
- **Set the scene:** Create a welcoming atmosphere (maybe some refreshments with herbs from your garden? Herbal lemonade?) and enough space for guests to explore.
- **Prepare some fun activities:**
    - Demonstrations of basic techniques
    - Tours of your setup
    - Q&A sessions
    - Maybe even a plant or seed swap!

Teaching Others: Passing It On

Remember how lost you felt at the beginning? Now's your chance to be the guide you wished you had! Whether it's:

- Mentoring a newbie gardener
- Writing a blog
- Creating social media content
- Leading workshops or classes
- Just sharing tips with curious neighbors, and answering questions

Teaching others doesn't just help them—it reinforces your own knowledge and often leads to new insights. Plus, there's nothing quite like seeing someone's eyes light up when their first seed sprouts!

SETTING NEW GOALS: KEEP GROWING!

Just like your plants are always reaching for the sky, keep stretching your own limits too. What's next in your gardening journey?

Goal Ideas to Consider:

- Master a new growing technique. (How bout them hydroponics?)

- Try exotic plant varieties.
- Create a more sustainable setup, aim for zero-waste.
- Start a community project.
- Become that neighborhood garden teacher.
- Write about your experiences.
- Teach others your favorite tricks.

Remember to make your goals SMART:

- Specific (not just "grow better plants")
- Measurable (how will you know you've succeeded?)
- Achievable (baby steps still move you forward!)
- Relevant (to what matters to you, aligns with your passions)
- Time-bound (but flexible, like a healthy vine)

CHECK-IN

Well, well. While you were busy coaxing seedlings and battling aphids, something remarkable happened. You transformed. Like a perennial coming back stronger each season, you've developed roots, gained resilience, and now you're blooming in ways you never imagined.

Through this chapter, you've discovered:

- The value of reflection and journaling to capture your greenhouse story.
- How your greenhouse can become a personal sanctuary for the soul.
- Ways to connect with and contribute to the gardening community.
- How to set goals for future growth.

Let's check on some things to see where you're at. Have you:

- Documented your greenhouse journey in some way?
- Found your own special way to enjoy your greenhouse as a retreat?
- Connected with other gardeners to share knowledge and experiences?
- Tried explaining basic greenhouse concepts to others?
- Grown anything from seed (or seedling) to harvest?

If you answered yes to even one of these, give yourself a pat on the back—you've come a long way! For the other ones you have yet to check off, you've got some great things to shoot for.

Before you continue, consider:

- Setting one meaningful "next step" for your greenhouse adventure.
- Finding one way to share your new knowledge with someone else.
- Joining one online community to connect with other gardeners.

Quick success tip: Start a "Victory Jar" in your greenhouse—a simple glass container where you drop in small notes about wins both leafy and personal. "First tomato ripened!" sits happily next to "Finally fixed that leaky valve!" When gardening gets tough, dip into your jar for an instant motivation boost that proves you're more capable than you think!

Add to that a gratitude practice, where you mentally name three things you're thankful for each time you enter your greenhouse. Maybe it's the warbler singing outside, those stubborn seeds that finally sprouted, or simply the quiet moment away from your phone. This tiny habit transforms routine maintenance into mindful moments that nourish both plants and soul.

Write these down and make a second jar to reach into, so you can always walk into your greenhouse and harvest something sweet, even when the plants are not yet ripe.

Final Wrap-Up

Standing in your greenhouse, surrounded by the life you've nurtured, it's hard not to see the parallels between gardening and life itself. Each seed you plant is potential waiting to unfold. Every challenge you overcome makes you more resilient. The connections you build, like roots spreading through soil, create a network of support and shared knowledge.

I've said it a bunch, I know, but it's so true it bears repeating. On this journey, you're not just growing plants; you're growing you. And ultimately, it is about building community and creating something beautiful that extends far beyond four walls and a roof. Whether you're journaling about your latest gardening triumph, sharing surplus tomatoes with neighbors, or teaching someone how to start their first seeds, you're contributing to something bigger than yourself.

As we wrap up our greenhouse adventure together, remember that every expert gardener started as a beginner who just kept growing. Your greenhouse is more than a place for plants—it's a space for learning, connecting, and blooming into your fullest potential.

Take a moment to appreciate where you are right now. You've created something amazing—a haven where both plants and people can thrive, and where all kinds of growth are nurtured.

And that's pretty dang cool.

CONCLUSION: YOU DID IT!

(OR AT LEAST, YOU'RE WELL ON YOUR WAY)

Well, look at you! You made it to the end of this book, and I bet you've got some stories to tell. Maybe a few triumphant harvests, definitely some "learning experiences" (that's what we call those mysterious plant disappearances in polite company), and probably at least one moment where you stared at a wilting seedling and wondered if plants could file for emotional distress.

Welcome to the club! Those first few plant casualties? They're like burnt cookies in the journey to becoming a master baker—practically a rite of passage. In fact, if you haven't killed at least one plant yet, are you even trying? (Though if you genuinely haven't, well ... please accept my enthusiastic congratulations while I quietly reevaluate everything I thought I knew about beginner gardening!)

Remember when you first cracked open these pages, maybe feeling a bit like I did when I started—staring at greenhouse options with the same bewildered expression my dog gives me when I try to explain why he can't eat chocolate? And now look at you! You're throwing around terms like "thermal mass" and "succession planting" like you've been speaking Plantese all your life.

You've learned that a greenhouse isn't just a fancy glass box for your tomatoes (though it is that too). It's your own perpetual slice of spring and summer, your timeout corner from the world, your "nobody bugs me when I'm gardening" sanctuary. Well, except for actual bugs. But hey, now you know how to deal with those too!

Speaking of dealing with things, let's take a moment to celebrate all you've conquered. You've mastered the art of climate control (aka learning how to checkmate nature), decoded the secret language of plants (from yellowing leaves to reddening tomatoes), and maybe even grown something exotic enough to make your neighbors wonder if you've secretly installed a teleporter to the tropics behind your potting bench.

But more than just growing plants, you've grown yourself. Every wilted leaf taught you patience. Every successful harvest built your confidence. And every time you figured out why your cucumber was throwing a tantrum, you added another tool to your mental gardening toolbox and built your resourcefulness.

Remember: gardening isn't about being perfect. It's about being persistent. Sometimes your plants will thrive despite your best efforts to pamper them, and sometimes they'll droop just to spite your master's degree in watering techniques. That's not failure—that's just keeping you humble. Besides, every "oops" moment is really just a future funny story waiting to be told, and wisdom to pass along, in seed form.

So what's next for you, greenhouse warrior? Maybe you'll experiment with those exotic herbs you can never find at the store. Or perhaps you'll become that friend who gives everyone homegrown tomatoes for Christmas. Whatever you choose to grow, you've got this.

And when you don't got this? Well, that's what your fellow gardeners are for. Don't be shy about joining local gardening groups or online forums. Gardeners are generally a friendly bunch—probably something to do with all that fresh air and vitamin D. Plus, we all love sharing our successes and commiserating over our spectacular failures. Nothing brings people together quite like swapping stories about the time we accidentally grew a zucchini so big we had to file it as a dependent on our taxes.

Before you close this book and rush off to implement your next brilliant garden scheme, I want to thank you. Thank you for letting me be your guide on this journey, for tolerating my plant puns, and for being brave enough to dream of growing your own little paradise. I know I'm sort of just disembodied words on a paper to you right now, but I want you to know I'm

CONCLUSION: YOU DID IT!

rooting for you with each seed you sow and each crop you bring in. And I couldn't be prouder of you.

You're part of something bigger now—a community of people who believe in the magic of putting a seed in soil and watching it become something amazing. We're all in this great big garden together.

Remember: your greenhouse journey is just beginning. There will be triumphs and tribulations, moments of pure joy, and times when you wonder if your plants are secretly plotting against you. You'll hit glorious strides when you feel like you've got this whole gardening thing figured out. Then, your plants will always find new ways to keep you guessing.

But remember, that's half the fun! Dancing the dance of life and figuring things out, and becoming better in the process. *Embrace it all.* Every wilted leaf and every successful harvest is part of your story.

Now go forth and grow something amazing! And if anyone asks where you learned all this? Tell them a friend named Jaden Rivers helped you out, and tell them where to grab a book.

Happy growing, my friend. May your greenhouse always be just the right temperature, your soil rich and crumbly, and your harvests bountiful.

A NOTE FROM THE PUBLISHER

Hey there, reader friend!

Thank you for spending time with this book—we hope it gave you the know-how, encouragement, and maybe even a few laughs at our mistakes that you needed to take your next step toward self-reliance. Around here, we believe in growing things that matter: good food, communities, and lives that feel a little more rooted and a lot more abundant.

If you found this book helpful, we'd be so grateful if you'd consider leaving a quick review. As a small, family-run publisher, your feedback means the world to us. It's how our books get into the next reader's hands, and it's how we stay visible, connected, and growing right alongside you.

We'd also love to invite you to join our reader team! It's free, and it means you'll get first dibs on new digital materials and sneak peeks before anyone else. (This is our way of handing you the VIP pass to the garden gate.)

And if you're itching to keep learning, we've got more books where this came from—covering a lot of ground from permaculture to prepping to homesteading, and more to come!

At All We Need Publishing, our motto is "we grow together." Thank you for being the best part of that, and being the tomato to our basil.

With appreciation,

—*The All We Need Publishing Team*

Free Stuff!

Wanna join our super-cool reader team and get access to free stuff? Stuff like our fresh-off-the-press ebooks, new audiobooks as we produce them, and other inspiring materials ...

If you're interested, just visit this website:

www.allweneedpublishing.com/readerteam

All We Need
PUBLISHING

amazon.com/author/allweneedpublishing
goodreads.com/allweneedpublishing
bookbub.com/profile/all-we-need-publishing
facebook.com/allweneedpublishing
instagram.com/allweneedpublishing
pinterest.com/allweneedpublishing

MORE FROM ALL WE NEED PUBLISHING

Permaculture Gardening for the Absolute Beginner

REFERENCES

247Garden. (2024, August 17). *How long will schedule-40 PVC pipes last under the sun?* https://www.247garden.com/blog/2024/08/17/how-long-will-schedule-40-pvc-pipes-last-under-the-sun/

Action Resource Management. (2024, September 27). *Greenhouse safety tips.* https://actionresourcemgmt.com/greenhouse-safety-tips/

Agran, H. (2024, April 17). *Step-by-step guide to canning.* Midwest Living. https://www.midwestliving.com/food/step-by-step-guide-to-canning/

Alterman, T. (2023, December 7). *How does a seed exchange work?* Mother Earth News. https://www.motherearthnews.com/organic-gardening/how-to-organize-a-community-plant-and-seed-swap/

Amerlife. (2024, August 23). *Top tips for an efficient greenhouse layout.* https://amerlifehome.com/blogs/news/top-tips-for-an-efficient-greenhouse-layout

Aquor Water Systems. (2024, December 18). *Top benefits of using a greenhouse for year-round gardening.* https://www.aquorwatersystems.com/blogs/news/greenhouse-benefits

Arcadia GlassHouse. (2017, February 10). *Tip #17: Vertical gardening in a greenhouse.* https://arcadiaglasshouse.com/greenhouse-tips/tip-17-vertical-gardening-greenhouse/

Baldwin, K. R., & Greenfield, J. T. (2024, May 9). *Composting on organic farms* (Publication No. AG-659-01). NC State Extension. https://content.ces.ncsu.edu/composting-on-organic-farms

Balzer, D. (2022, January 5). *The best soil for greenhouse gardening: A comprehensive guide.* BC Greenhouse Builders Ltd. https://blog.bcgreenhouses.com/greenhouse-garden-tips-the-best-soil-primer

Bartok, J. W., Jr. (2003). *Reducing humidity in the greenhouse.* University of Massachusetts Amherst Center for Agriculture, Food, and the Environment. https://ag.umass.edu/greenhouse-floriculture/fact-sheets/reducing-humidity-in-greenhouse

Bartok, J. W., Jr. (2005). *Ventilation for greenhouses.* University of Massachusetts Amherst Center for Agriculture, Food, and the Environment. https://ag.umass.edu/greenhouse-floriculture/fact-sheets/ventilation-for-greenhouses

Bartok, J. W., Jr. (2013). *Design and layout of a small commercial greenhouse operation.* University of Massachusetts Amherst Center for Agriculture, Food, and the Environment. https://ag.umass.edu/greenhouse-floriculture/fact-sheets/design-layout-of-small-commercial-greenhouse-operation

BBC Gardeners' World Magazine. (2022, February 24). *Prepare your greenhouse for spring.* https://www.gardenersworld.com/plants/prepare-your-greenhouse-for-spring/

Bessin, R., Townsend, L. H., & Anderson, R. G. (2007). *Greenhouse insect management* (Publication No. ENT-60). University of Kentucky Cooperative Extension Service. https://entomology.ca.uky.edu/ent60

Blalock, A. (n.d.). *Identifying and managing plant disorders in greenhouse production.* Tennessee State University. https://www.tnstate.edu/faculty/ablalock/documents/Identifying%20and%20Managing%20Plant%20Disorders%20in%20Greenhouses%20Production1.pdf

REFERENCES

Bruno, A. (2024, January 9). *How much does a greenhouse cost in 2025?* Angi. https://www.angi.com/articles/how-much-do-greenhouses-cost.htm

Bull, R. (2025, January 8). *How to set your gardening goals – for a joyous and fruitful yard in 2025.* Homes & Gardens. https://www.homesandgardens.com/gardens/how-to-set-gardening-goals

Bundy, H., & Bradley, L. (2024, September 22). *Gardening as a mindfulness practice.* NC State Extension. https://extensiongardener.ces.ncsu.edu/2024/07/gardening-as-a-mindfulness-practice/

Cacho, J. (2019, February 13). *Advantages of drip irrigation in greenhouses.* Mundoriego. https://mundoriego.es/en/advantages-of-drip-irrigation-in-greenhouses/

CERES Greenhouse Solutions. (2021, January 24). *Year-round greenhouse planting calendar.* https://ceresgs.com/year-round-greenhouse-planting-calendar/

Cloyd, R. A., Jandricic, S., & Lindberg, H. (2024, January). *Commercially available biological control agents for greenhouse insect and mite pests* (Extension Bulletin E-3299). Michigan State University Extension. https://www.canr.msu.edu/floriculture/uploads/files/E3299_COMMERCIALLY_AVAILABLE_BIOLOGICAL_2024.pdf

Crane, J. H. (2016). *Pineapple growing in the Florida home landscape* (Publication No. MG055). University of Florida Institute of Food and Agricultural Sciences. https://doi.org/10.32473/edis-mg055-2005

Dave's Garden. (n.d.). *Common greenhouse problems and solutions.* Retrieved April 18, 2025, from https://davesgarden.com/guides/articles/common-greenhouse-problems-and-solutions

de Beer, E. (2023, July 11). *What is the difference between LED and HPS grow lights?* Philips Lighting. https://www.usa.lighting.philips.com/application-areas/specialist-applications/horticulture/hortiblog/light-and-growth/what-is-the-difference-between-led-and-hps-grow-lights

Dengarden. (2023, March 24). *Top 10 gardening forums, chat rooms, and communities.* https://dengarden.com/gardening/10-online-gardening-forums-for-green-thumbs

Deziel, C. (2024, November 15). *9 types of greenhouses.* Family Handyman. https://www.familyhandyman.com/article/types-of-greenhouses/

Diver, S. (2001, January). *Compost heated greenhouses* (Publication No. CT071). Appropriate Technology Transfer for Rural Areas. https://fyi.extension.wisc.edu/energy/files/2016/09/compostheatedgh.pdf

Duncan, L. (2022, October 4). *Was your power out? Check indoor humidity.* UF/IFAS Extension Sumter County Blog. https://blogs.ifas.ufl.edu/sumterco/2022/10/04/check-indoor-humidity/

ECOgardener. (2018, November 18). *The pros and cons of organic and chemical fertilizers.* https://ecogardener.com/blogs/news/the-pros-and-cons-of-organic-and-chemical-fertilizers

The Editors of Organic Life. (2017, December 18). *The 7 best organic pest control techniques for your garden.* Good Housekeeping. https://www.goodhousekeeping.com/home/gardening/a20705693/organic-pest-control/

Farmer, T. (2023, September 11). *How much does a greenhouse cost?* HomeGuide. https://homeguide.com/costs/greenhouse-cost

Fisher, S. (2025, February 24). *12 free DIY greenhouse plans.* The Spruce. https://www.thespruce.com/free-greenhouse-plans-1357126

Fresh Farms. (2024, February 8). *How to pick perfectly ripe produce at the grocery store.* https://www.freshfarms.com/how-to-pick-perfectly-ripe-produce-at-the-grocery-store/

Gardening Express. (2023, August 9). *Growing exotic plants in greenhouses.* Gardening Express

REFERENCES

Knowledge Hub. https://help.gardeningexpress.co.uk/knowledge-base/growing-exotic-plants-in-greenhouses/

Gast, K. L. B. (2001). *Storage conditions: Fruits and vegetables* (Bulletin No. 4135e). University of Maine Cooperative Extension. https://extension.umaine.edu/publications/4135e/

Grant, A. (2023, January 9). *Greenhouse location guide: Learn where to put your greenhouse.* Gardening Know How. https://www.gardeningknowhow.com/special/greenhouses/where-to-put-greenhouse.htm

Hansen, B. J. (n.d.). *Why keeping a garden journal is important.* GardenTech. Retrieved April 18, 2025, from https://www.gardentech.com/blog/gardening-and-healthy-living/start-a-garden-journal-this-season

Harding, J. (2024, October 2). *The best harvesting tools for this year's garden bounty.* Bob Vila. https://www.bobvila.com/lawn-and-garden/best-harvesting-tools/

Hellmann, M. (2021, February 23). *Simple checklist for maintaining your greenhouse.* Hummert International. https://www.hummertinternational.com/tip/greenhouse-maintenance/

Hortinergy. (n.d.). *Hortinergy - Online greenhouse design software.* Retrieved April 18, 2025, from https://www.hortinergy.com/

How To Garden Well. (2024, October 22). *How to create a budget-friendly greenhouse for beginners.* https://howtogardenwell.com/how-to-create-a-budget-friendly-greenhouse-for-beginners/

INSONGREEN. (2023, December 31). *Best temperature monitoring system for your greenhouse.* https://www.insongreen.com/greenhouse-temperature-monitoring-system/

International Plastics, Inc. (2022, October 31). *Greenhouse building materials: Should I choose glass, polycarbonate, or poly film?* https://www.interplas.com/blog/materials-to-build-a-green-house-diy

James, J. (2024, September 25). *42 easy to grow greenhouse plants for beginners.* Greenhouse Emporium. https://greenhouseemporium.com/easy-to-grow-greenhouse-plant/

Kapoor, L., Simkin, A. J., Doss, C. G. P., & Siva, R. (2022). Fruit ripening: Dynamics and integrated analysis of carotenoids and anthocyanins. *BMC Plant Biology, 22*(1), Article 27. https://doi.org/10.1186/s12870-021-03411-w

Karlsson, M. (2024). *Controlling the greenhouse environment* (Publication No. HGA-00336). University of Alaska Fairbanks Cooperative Extension Service. https://www.uaf.edu/ces/publications/database/gardening/files/pdfs/HGA-00336-Controlling%20GreenHouse%2011-27-24.pdf

Kluepfel, M., Blake, J. H., Keinath, A. P., & Williamson, J. (2021, May 26). *Tomato diseases & disorders* (Factsheet HGIC 2217). Clemson Cooperative Extension Home & Garden Information Center. https://hgic.clemson.edu/factsheet/tomato-diseases-disorders/

Lofgren, K. (2023, November 13). *11 essential greenhouse supplies to get started.* Gardener's Path. https://gardenerspath.com/how-to/greenhouses-and-coldframes/greenhouse-supplies/

Masabni, J., King, S., & Proctor, N. (2013). *Easy gardening: Watering your vegetables* (Publication No. EHT-024). Texas A&M AgriLife Extension. https://aggie-horticulture.tamu.edu/vegetable/wp-content/uploads/sites/10/2013/09/eht_024_watering_your_vegetables.pdf

McLaughlin, C. (2009, June 12). *The benefits of joining a community garden.* Fine Gardening. https://www.finegardening.com/article/the-benefits-of-joining-a-community-garden

Millcreek Gardens. (2018, July 23). *5 common gardening myths debunked.* https://millcreekgardens.com/5-common-gardening-myths-debunked/

Mud Hub Greenhouses. (2023, July 20). *The most common greenhouse gardening mistakes and how to avoid them.* https://mudhubgreenhouses.com/the-most-common-greenhouse-gardening-mistakes-and-how-to-avoid-them/

REFERENCES

National Parks Board. (n.d.). *Basic gardening skills.* Retrieved April 18, 2025, from https://gardeningsg.nparks.gov.sg/new-to-gardening/basic-gardening-skills/

New Hampshire Agricultural Experiment Station. (n.d.). *MacFarlane Research Greenhouses open house.* Retrieved April 18, 2025, from https://colsa.unh.edu/new-hampshire-agricultural-experiment-station/news-events/macfarlane-research-greenhouses-open-house

NSW Department of Primary Industries. (n.d.). *Greenhouse covering materials.* Retrieved April 18, 2025, from https://www.dpi.nsw.gov.au/agriculture/horticulture/greenhouse/structures-and-technology/covers

Oleszak, H., Hammond, E., & Murgel, J. (2024, May 6). *Choosing a soil amendment* (Fact Sheet No. 7.235). Colorado State University Extension. https://extension.colostate.edu/topic-areas/yard-garden/choosing-a-soil-amendment/

Pickens, J., & Wells, D. (2022, September 30). *Controlling the greenhouse environment for vegetable crops.* Alabama Cooperative Extension System. https://www.aces.edu/blog/topics/crop-production/controlling-the-greenhouse-environment-for-vegetable-crops/

Planta Greenhouses. (2023, November 27). *Energy-smart gardening: 19 strategies for heating & cooling your DIY backyard greenhouse.* https://plantagreenhouses.com/blogs/learn/energy-smart-gardening-19-strategies-for-heating-and-cooling-your-diy-backyard-greenhouse

Planta Greenhouses. (2024, January 2). *Types of greenhouse bases: With pros & cons of each.* https://plantagreenhouses.com/blogs/greenhouse-bases/types-of-greenhouse-bases-with-pros-cons-of-each

Pochubay, E., Himmelein, J., Elzinga, M., & Grieshop, M. J. (n.d.). *Companion plants in greenhouses: Potential for pest monitoring, trapping, and natural enemy open rearing.* Michigan State University Extension. https://www.canr.msu.edu/foodsystems/uploads/files/companion-plants-in-greenhouses.pdf

Sears, C. (2022, March 11). *How to use a moisture meter for your plants.* The Spruce. https://www.thespruce.com/how-to-use-a-moisture-meter-5220649

Siggins, R. (2022, September 2). *How to build a greenhouse with old windows and doors.* SoCo Wood & Windows Blog. https://socoww.com/contractor/how-to-build-a-greenhouse-with-old-windows-and-doors/

Sparks, B. (2019, April 25). *8 environmental controls to help you monitor your greenhouse environment.* Greenhouse Grower. https://www.greenhousegrower.com/technology/7-environmental-controls-to-help-you-monitor-your-greenhouse-environment/

Sun Direction. (n.d.). *Sun direction: Orientation of the sun throughout the day.* Retrieved April 18, 2025, from https://sun-direction.com/

Talerico, D. (2019, October 18). *A beginner's guide to using a hobby greenhouse.* Homestead and Chill. https://homesteadandchill.com/beginners-guide-using-hobby-greenhouse/

Texas A&M AgriLife Extension. (n.d.-a). *Greenhouse vegetables (Cucumber, tomato, lettuce) crop guide.* Retrieved April 18, 2025, from https://aggie-horticulture.tamu.edu/smallacreage/crops-guides/greenhouse-nursery/greenhouse-vegetables/

Texas A&M AgriLife Extension. (n.d.-b). *Irrigating greenhouse crops.* Retrieved April 18, 2025, from https://aggie-horticulture.tamu.edu/ornamental/greenhouse-management/irrigating-greenhouse-crops/

U.S. Environmental Protection Agency. (2024, September 3). *Integrated Pest Management (IPM) principles.* https://www.epa.gov/safepestcontrol/integrated-pest-management-ipm-principles

University of California Agriculture and Natural Resources Marin Master Gardeners. (n.d.). *Garden myths busted.* Retrieved April 18, 2025, from https://marinmg.ucanr.edu/BASICS/GARDEN_MYTHS_BUSTED/

REFERENCES

University of Michigan. (2003). *A guide for preparing soil profile descriptions.* https://websites.umich.edu/~nre430/PDF/Soil_Profile_Descriptions.pdf

Walker, S., & Joukhadar, I. (2019). *Greenhouse vegetable production* (Circular 556). New Mexico State University Cooperative Extension Service. https://pubs.nmsu.edu/_circulars/CR556/index.html

Washington State University Extension Clark County. (n.d.). *Greenhouse structures.* Retrieved April 18, 2025, from https://extension.wsu.edu/clark/agbusiness/garden-center-nursery-management/greenhouse-structures/

Weisenhorn, J., & Hoidal, N. (2024). *Lighting for indoor plants and starting seeds.* University of Minnesota Extension. https://extension.umn.edu/planting-and-growing-guides/lighting-indoor-plants

Wilson, C., & Bauer, M. (2014). *Drip irrigation for home gardens* (Fact Sheet No. 4.702). Colorado State University Extension. https://extension.colostate.edu/topic-areas/yard-garden/drip-irrigation-home-gardens-4-702/

Zac, J. (2025, January 1). *Is it illegal to collect rainwater: 2025 complete state guide.* World Water Reserve. https://worldwaterreserve.com/is-it-illegal-to-collect-rainwater/

Printed in Dunstable, United Kingdom